I0810032

IMAGES
of America

CONNEAUT

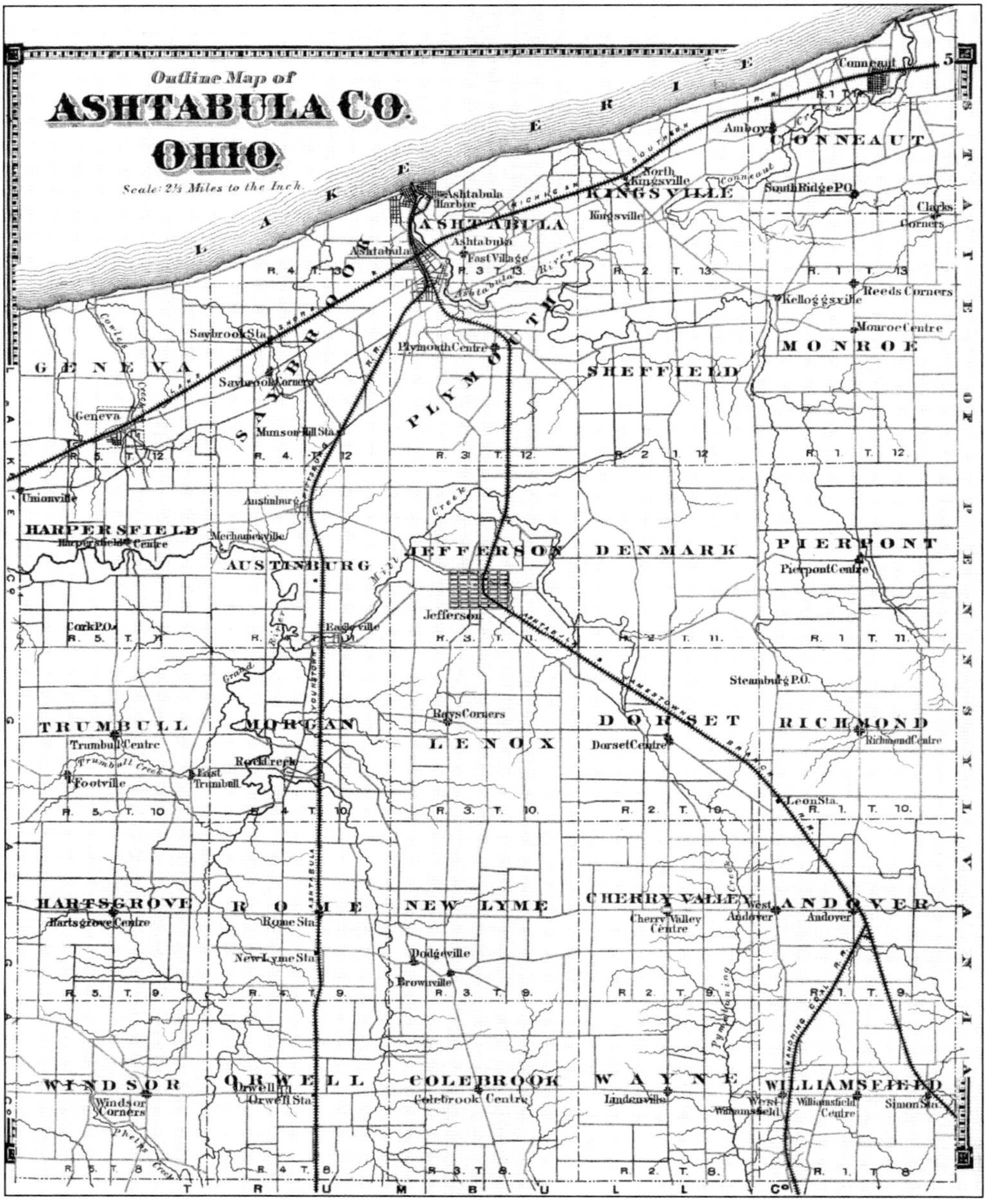

This 1874 map shows the 28 townships of Ashtabula County, Ohio. When the Moses Cleaveland party surveyed the Western Reserve, the county was laid out in 5-mile-square townships, with the exception of the ones bordering on Lake Erie. The primary focus of the book is on the city of Conneaut and Conneaut Township. The last chapter features Kingsville Township, Monroe Township, and North Springfield, Pennsylvania. (Grand Valley Public Library.)

On the Cover: A total of 43 of these steamships, designed by Capt. Alexander McDougall, were built from 1887 to 1898. They were called "whalebacks" because only the rounded portion of the hull was seen above the waterline when fully laden. Since the bow bears some resemblance to a pig's snout, another derisive term it acquired was "pig boat." Due to their frequent presence in Conneaut Harbor, the area became known as "Pig Harbor." (Conneaut Area Historical Society.)

IMAGES
of America

CONNEAUT

David B. Owens in association with the
Conneaut Area Historical Society

ARCADIA
PUBLISHING

ISBN 978-0-7385-7731-9

Published by Arcadia Publishing
Charleston, South Carolina

Printed in the United States of America

Library of Congress Control Number: 2009929496

For all general information contact Arcadia Publishing at:
Telephone 843-853-2070
Fax 843-853-0044
E-mail sales@arcadiapublishing.com
For customer service and orders:
Toll-Free 1-888-313-2665

Visit us on the Internet at www.arcadiapublishing.com

In memory of Conneaut's "Hardware Lady," Doris Bender Owens (January 27, 1922 to November 22, 2008)

Contents

ACKNOWLEDGMENTS

My mother, Doris Owens, worked for several years in the Harbor Hardware store at 985 Broad Street. She asked me if I could determine when the building was constructed for the current owner of the building and Harbor Beverage, Rob Harkabus. This investigation ultimately led me to contact Ken Springirth, the author of six books published by Arcadia Publishing.

During my first contact with Ken, we spent a great deal of time discussing Conneaut's trolleys and its history in general. Several more conversations took place, culminating with Ken encouraging me to write a book on the history of Conneaut and having it published by Arcadia Publishing. It is because of my mother's love for Conneaut, and the above series of curious events, that I undertook the writing of this book.

Through my research, I have renewed friendships with people I have not seen in several decades. Jack Morton, Karl and Pat (Haas) Rowbotham, and Joyce (Wright) Mackey have contributed their photographs, memories, and encouragement to ensure that this book depicts Conneaut through the eyes and experiences of many who share fond memories of growing up in Conneaut and hope for its bright future.

I have met many people during the past few months who have assisted in researching this book who I thus became friends with in the process. Jack Mathay, Sue and Arnie Howard, Jerry Gee, Norm Gross, Joan Barnett, Sally Richards, and many other members of the Conneaut Area Historical Society gave freely of their time, photographs, and knowledge to ensure the completeness and accuracy of this book. Andy Pochatko and his "History of Conneaut Ohio" Web site (conneautohio.tripod.com) have been sources of information and inspiration for me and countless others over the years.

I gained a wealth of knowledge of the surrounding communities from Mariana Branch, Joan Hockenberry, and Jennifer Krenisky.

My wife, Carol, endured months of listening to me talk about people from the 19th century as if they were still walking the streets of Conneaut. Despite all of this, she still had the patience to proofread the book.

Most of all, I would like to thank my Arcadia Publishing editor, Melissa Basilone, for having faith in me and this project.

Unless otherwise noted, all images appear courtesy of the Conneaut Area Historical Society.

INTRODUCTION

In May 1796, a party of Connecticut Land Company surveyors left Schenectady, New York, to survey the land beyond the western boundary of Pennsylvania. This territory was known as the Western Reserve. This band of 48 men and two women, under the command of Gen. Moses Cleaveland, sailed along Lake Erie arriving at the mouth of the Conneaut River on July 4, 1796. They named the location Port Independence.

The Iroquois tribe living here was the Massasagua, whose members survived by hunting, fishing, and primitive farming. They lived in huts roofed with bark strips until their departure in 1798.

By 1804, the settlement, which had grown up along the banks of the river, was organized as Salem. In 1834, the village incorporated and changed its name to Conneaut, meaning "river of many fish." When Conneaut incorporated, the population center and main business district was Conneaut Village, which was surrounded by Conneaut Township. The township communities became known as East Conneaut, South Ridge (Farnham), and Amboy.

In 1818, shipbuilding began in Conneaut with the 27-ton *Salem Packet* being launched south of the Main Street bridge. Faster and larger ships were built each year. The first government appropriation for harbor improvement, in 1829, resulted in another increase in shipbuilding.

In 1834, the first steamer built at Conneaut Harbor was the 300-ton *North America*. The largest ship ever built in Conneaut Harbor was a three-masted wooden schooner 174 feet in length that had a 39-foot beam and a 13-foot draft. With an 800-ton capacity, *Ogarita* was the largest ship on the Great Lakes when launched in 1864.

By 1835, the Old Plank Road was laid from the business section of town to the waterfront piers in support of the blossoming shipbuilding industry.

By 1837, Conneaut had a bank, printing office, steam saw mill, iron foundry, three churches, five flour mills, 11 taverns, and 24 business establishments. Eighty homes and several stores were in the process of being built. The harbor traffic consisted of 275 sailing vessels and 760 steamers. Outgoing cargo consisted of sawed lumber, pipe staves, grain, pork and beef barrels, glass, coal, cheese, butter, fruit, and cast iron. Incoming cargo was salt, pine lumber, gypsum, fish, limestone, and burr-mill stones.

The most prosperous shipbuilding years were from 1845 to 1852. In 1852, the Cleveland, Painesville and Ashtabula Railroad (later known as New York Central Railroad) reached Conneaut, and the harbor experienced a sharp decline in freight handling. With the decline in shipping and shipbuilding, the piers and harbor deteriorated rapidly, and Conneaut Harbor became merely a fishing port.

The main focus of this book is the nearly 100-year cycle from the 1860s through the 1960s. During this period, Conneaut transformed itself from a bucolic agricultural and fishing village to a major railroad, shipping, and manufacturing center. This book dovetails with Arcadia Publishing's *Railroading in Conneaut, Ohio*, *Greater Erie Trolleys*, and *Bessemer and Lake Erie Railroad*. Aspects of Conneaut not covered by these books are explored in greater depth. Conversely, railroading and trolleys are touched upon lightly.

Conneaut began to emerge as an industrial center following the Civil War because of its business leaders, who lobbied heavily to convince businesses to locate in Conneaut.

Like most small towns in America, Conneaut's prosperity cycle ended with the changing times. The interstate highway system and airplane travel supplanted the railroads as the preferred modes of transportation. Mergers and acquisitions replaced small business with conglomerates located in large cities. Foreign competition closed the Pittsburgh steel mills, which were Conneaut Harbor's lifeblood for nearly a century. As of this writing, the once proud 858-foot ore carrier *Roger Blough* sits rusting in Conneaut Harbor.

Historian David C. McCullough once said, "History is a guide to navigation in perilous times. History is who we are and why we are the way we are."

Conneaut's success during its past perilous times was brought about by people working together and reaching out to bring new opportunities to the community. Business leaders saw to it that Conneaut was easy to do business with. They were men and women of vision who took it upon themselves to see that businesses like the Nickel Plate Railroad and Pittsburgh and Conneaut Dock Company found Conneaut more attractive than its competitors.

Three railroads, CSX, Norfolk and Southern, and Bessemer and Lake Erie still serve the city today. The 70 acres, which the city gave to the Nickel Plate Railroad in 1881, lies dormant, crying out to once again be a major asset to the community. Additionally, the city has 10 miles of undeveloped frontage along U.S. Interstate 90.

Conneaut needs people with the vision to learn from the past and build a bright future with the resources and opportunities at hand. Conneaut still has advantages of a limitless supply of fresh water and its location as a port on Lake Erie.

Recently concerned citizens have taken action into their own hands. Emerging from a grassroots movement is the Citizens for Economic Development (CFED). In the late 19th century, leading citizens, such as Mayor Sagito J. Smith, personally assumed the responsibility for attracting new businesses. The 21st century approach, and the CFED recommendation, is for cities to employ an economic development director to attract and spur economic growth. In today's perilous times, perhaps an economic development director is exactly what Conneaut needs.

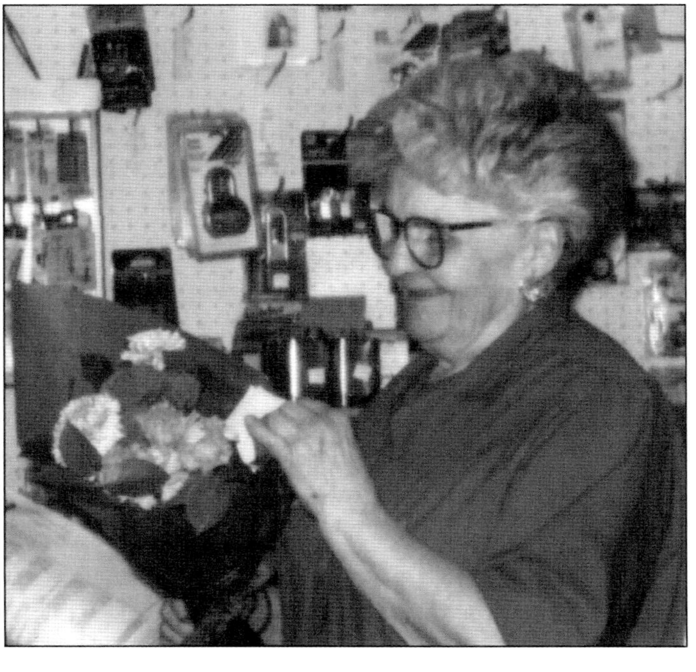

Doris Owens moved to Conneaut with her husband, Melvin, an employee of The Astatic Corporation for nearly 50 years, and son David in 1947. She loved Conneaut and was active in the community with her four children in scouting and the Conneaut High School Band Boosters. She roller-skated competitively and worked until her mid-70s. Combining 25 years at Harbor Hardware and Chapin's West End Hardware, she became known as Conneaut's "Hardware Lady."

One

Lake Erie

Native Americans harvested fish from Lake Erie using nets fashioned from twisted and knotted strands of willow bark. Beginning with the establishment of Port Independence by Moses Cleaveland and his surveying party, Lake Erie would also play an important role in the lives of the white settlers.

As the village of Salem (Conneaut) grew, ships began stopping at the harbor. On March 3, 1829, Congress appropriated $7,500 to improve the harbor. By the end of the season, an eastern pier measuring 360 feet in length and western pier of 270 feet were completed on either side of the Conneaut River. The piers consisted of timber cribs (base) filled with stone. The depth of the channel was increased from 2 feet to 12 feet.

In 1831, the scow schooner *Olive Branch* delivered the first cargo of iron ore, from the East, to the harbor. The ore was consigned to Ohio Furnace, located on Conneaut Creek near Furnace Road. The foundry, which employed 150 men, had exhausted the local supply of bog ore used in its production of cooking utensils, stoves, and tools.

By 1833, additional improvements were made, extending the piers and adding 120-foot legs that fanned out, leaving a 176-foot opening at the mouth of the harbor. The depth of the channel varied from 8 to 16 feet.

On May 19, 1835, Conneaut Harbor's first lighthouse was completed. The lighthouse tower was 40 feet tall from its crib to the center of the light. The crib rose 21 feet above the level of the lake and was constructed of stone. The lantern was 6 feet square and housed six reflectors 13 inches in diameter. The purpose of the reflectors was to amplify the light, which consisted of four simple kerosene lamps.

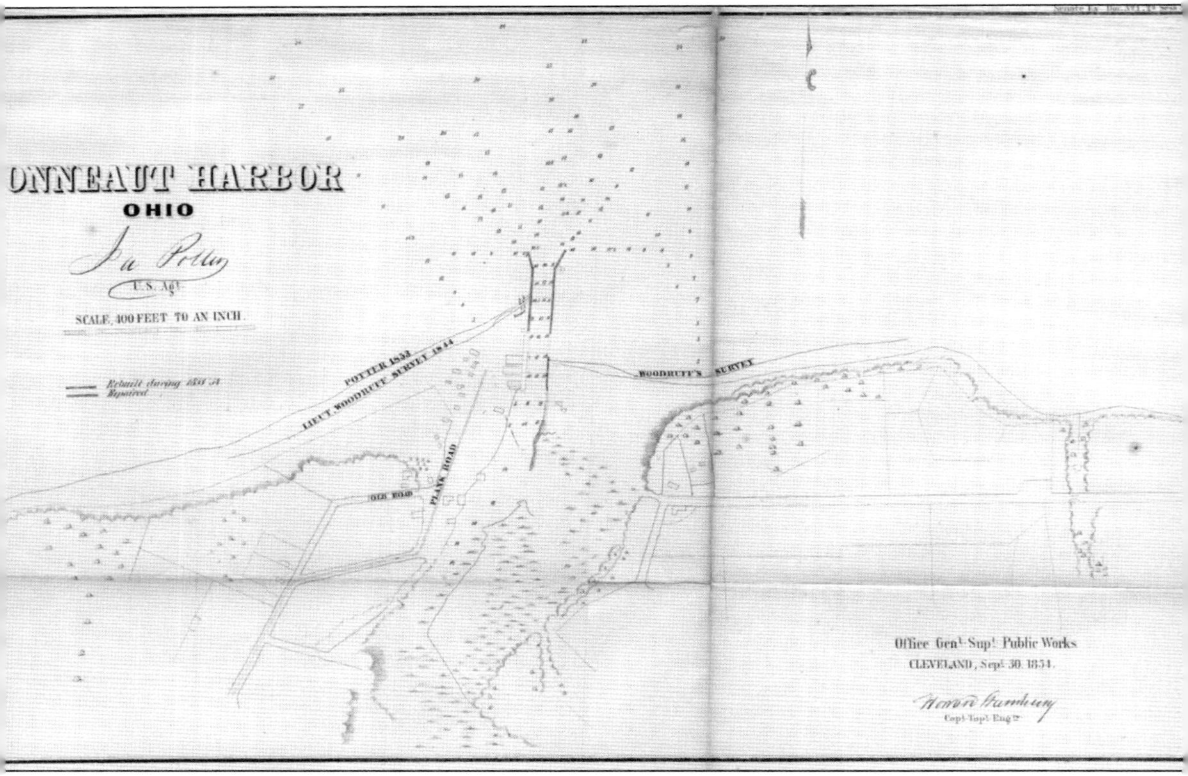

The location of the 1835 lighthouse was approximately 120 feet south of the eastern pier as indicated on this 1854 harbor chart. All lighthouses from Cleveland to Conneaut were built on the eastern side of the harbor to minimize the possibility of mishaps resulting from confusion when entering the ports.

By 1895, the west pier had been extended 300 feet. Four hundred feet of elevated walkway was rebuilt to provide a means of reaching the light during storms when waves washed completely over the pier. A new tower was erected at a height of 35 feet with a white beacon, which first glowed on September 8, 1895. In 1896, a project was authorized for the construction of two detached breakwalls.

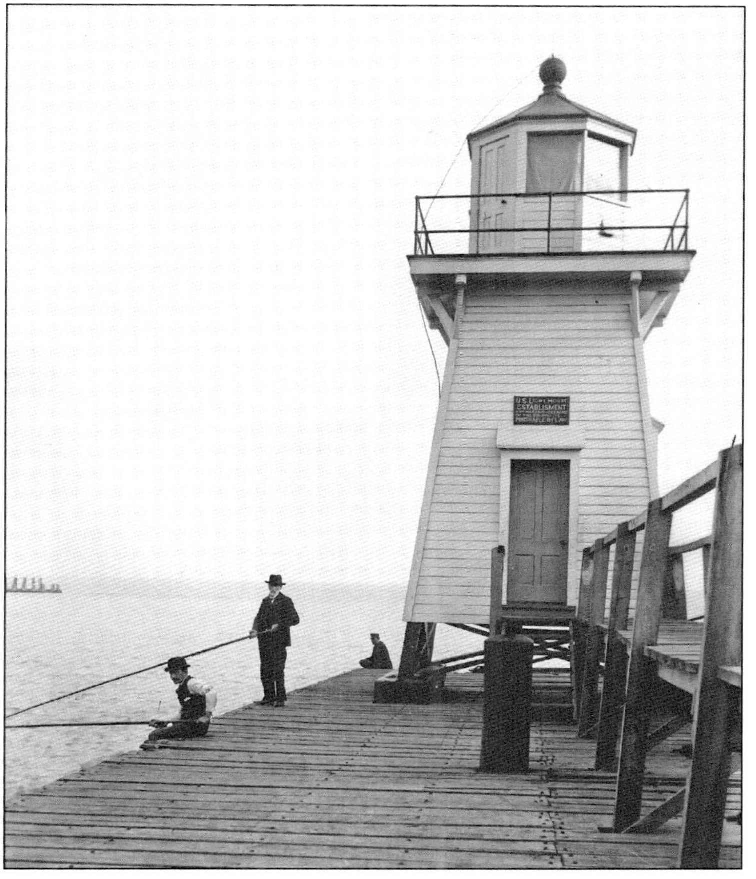

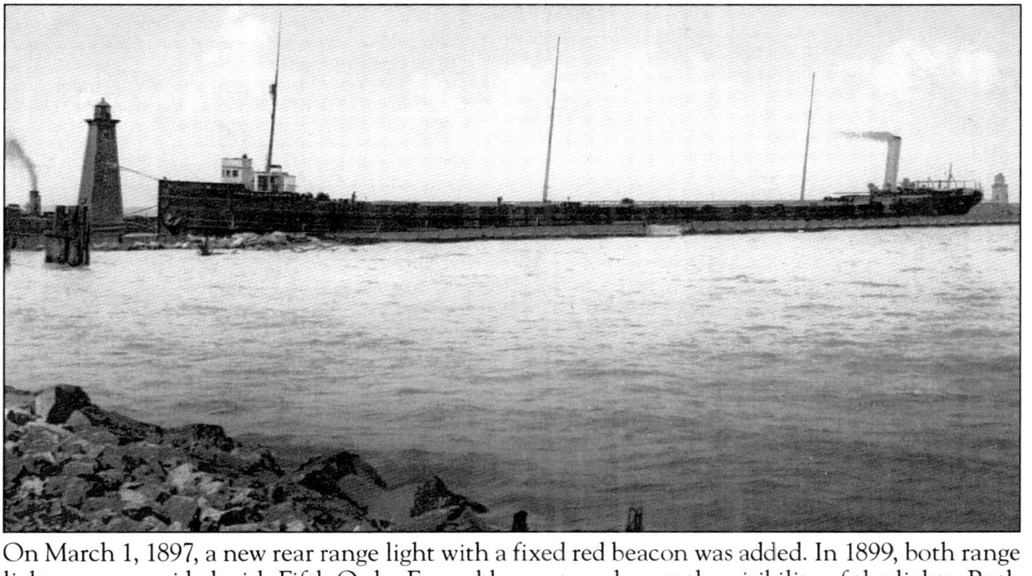

On March 1, 1897, a new rear range light with a fixed red beacon was added. In 1899, both range lights were provided with Fifth Order Fresnel lenses to enhance the visibility of the lights. Both lights can be seen in this 1900 postcard of the 478-foot-long *Isaac L. Ellwood.*

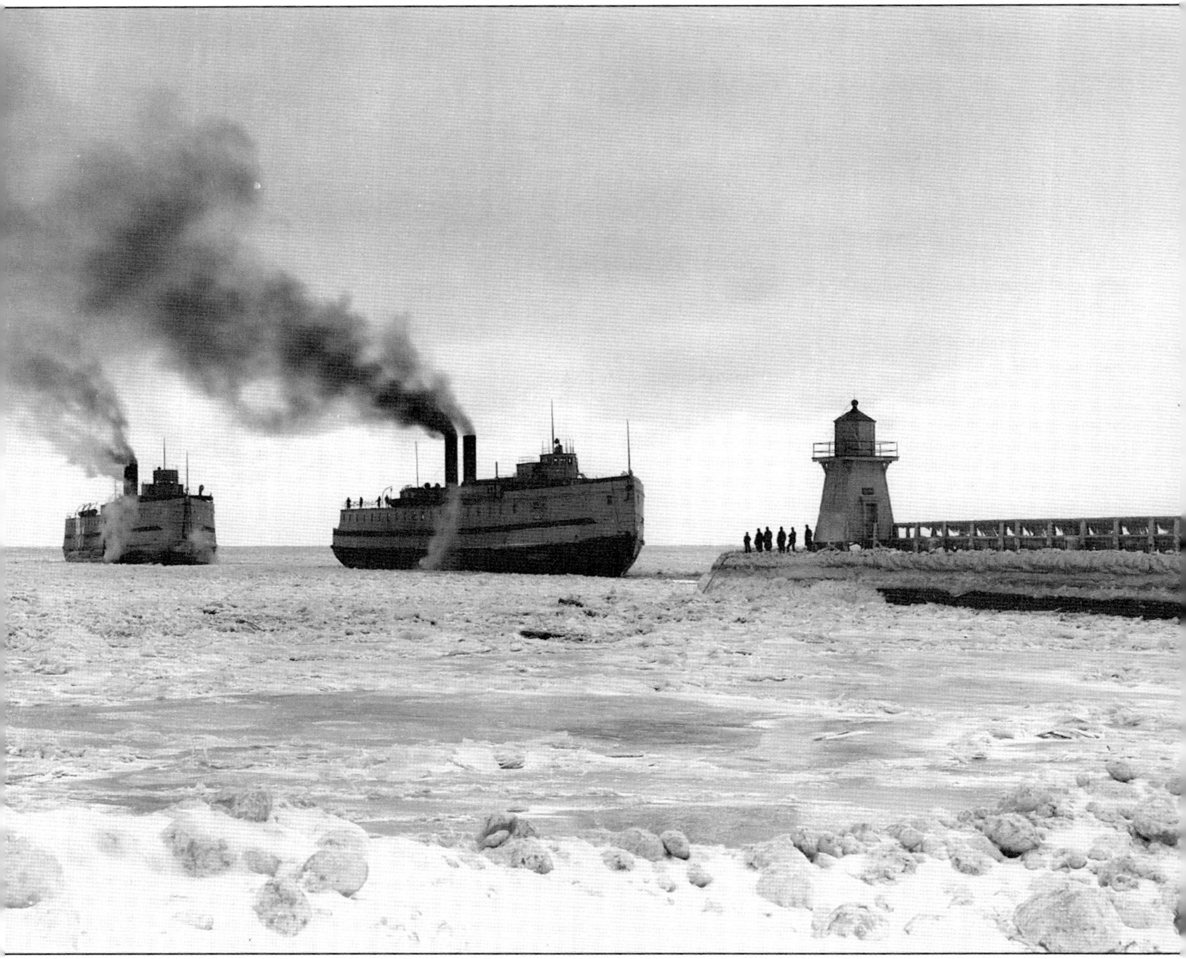

In this pre-1898 photograph, *Shenango I* uses her stern propellers to churn pack ice to clear a path for *Shenango II*. On January 8, 1904, the *Shenango I* became trapped in ice within sight of Conneaut Harbor. On March 11, she caught fire and burned until March 29, when she sank. The hull sat more than 20 feet below the surface until salvage operations were completed in June 1907.

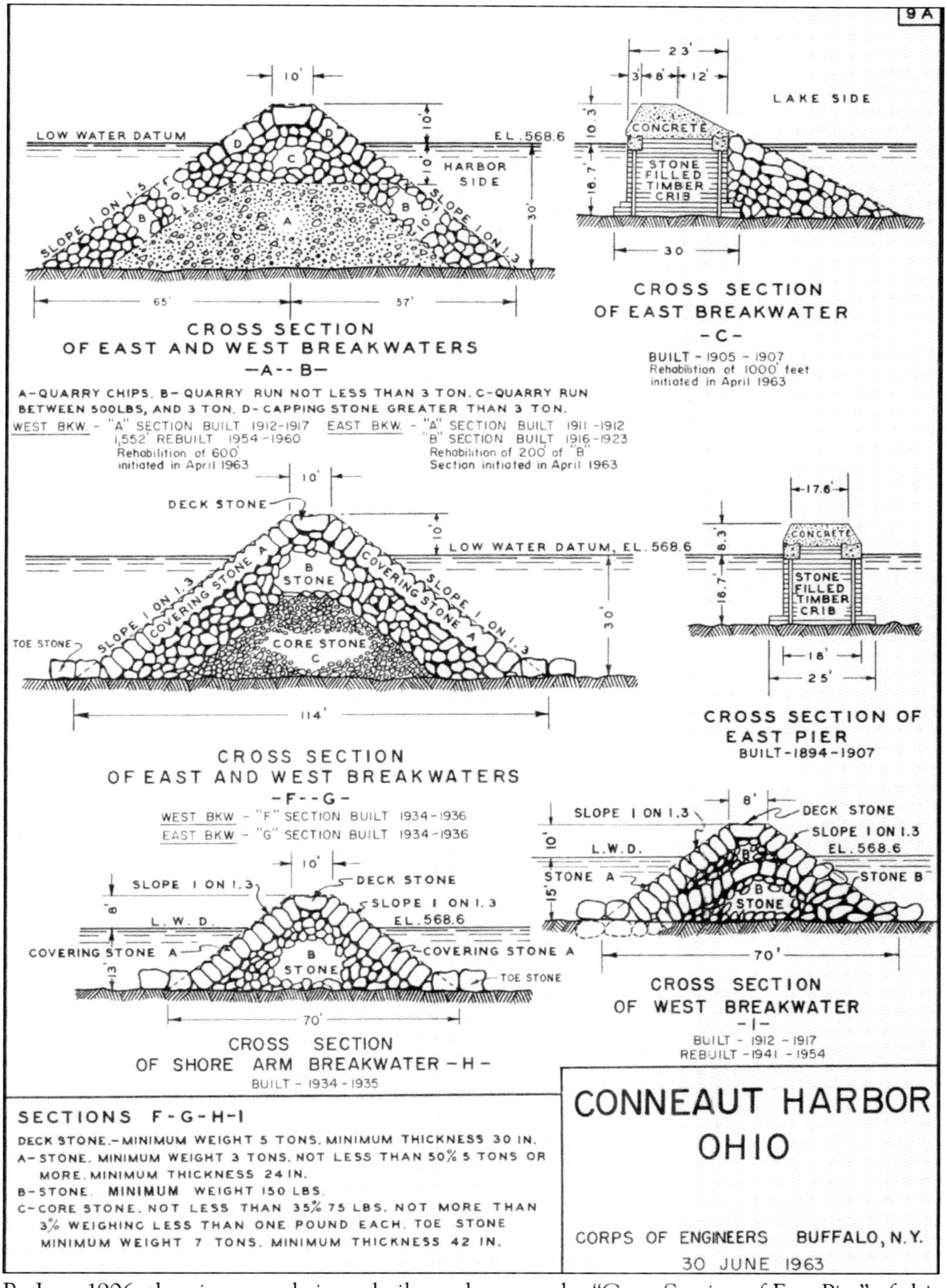

By June 1906, the piers were being rebuilt, as shown under "Cross Section of East Pier" of this image. These diagrams are included for a complete understanding of the evolution of pier and breakwall construction between 1894 and 1960. Although the term *breakwater* is used by the corps of engineers, the Great Lakes colloquial term is *breakwall*. (U.S. Army Corps of Engineers.)

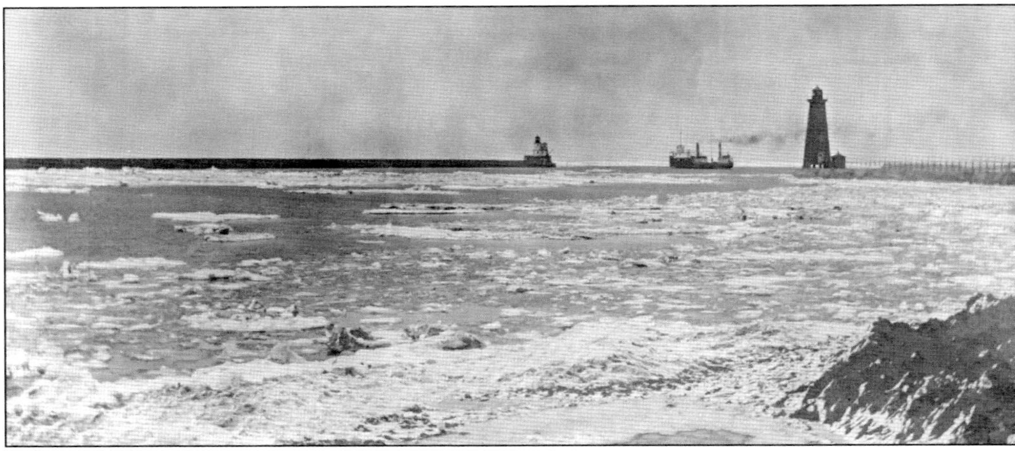

In 1906, new front and rear range lights were completed and lighted for the first time on October 13. The rear range light (right) sat at the end of the newly rebuilt west pier of Conneaut Harbor. This light had a fixed red signal, which could be seen for 9.75 miles.

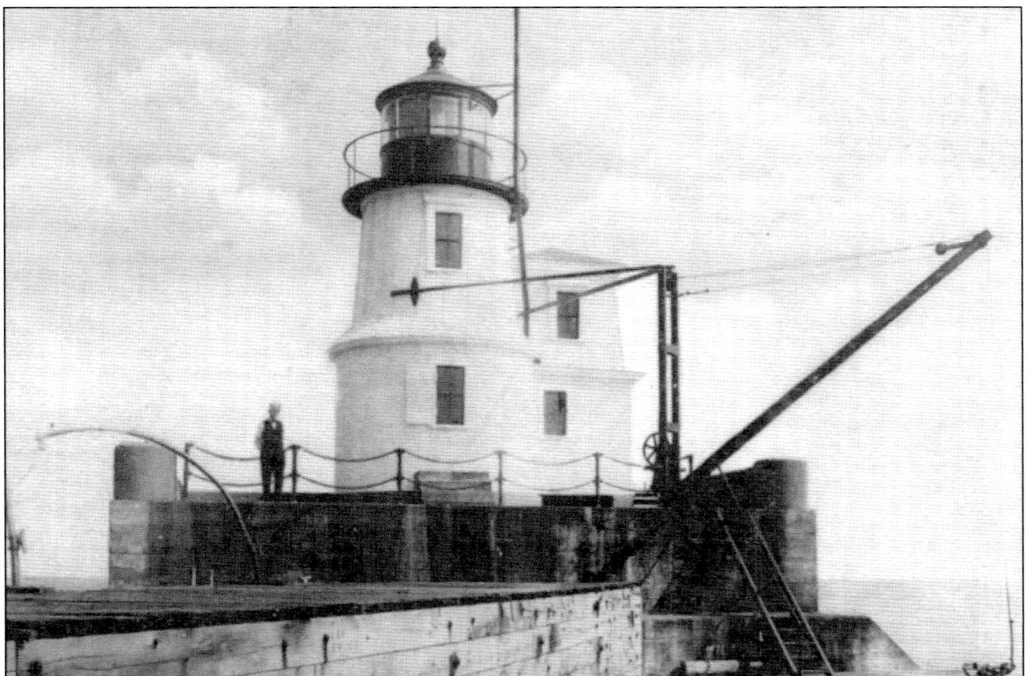

The new front range light was completed at the northern end of the new west breakwall. The new front-range beacon was a fixed white signal that could be seen for 13.75 miles. A mechanical fog bell was added in 1910. The 1,054-foot east breakwall was begun in 1905 and completed in 1907. From 1911 to 1912, the east breakwall was extended an additional 810 feet shoreward.

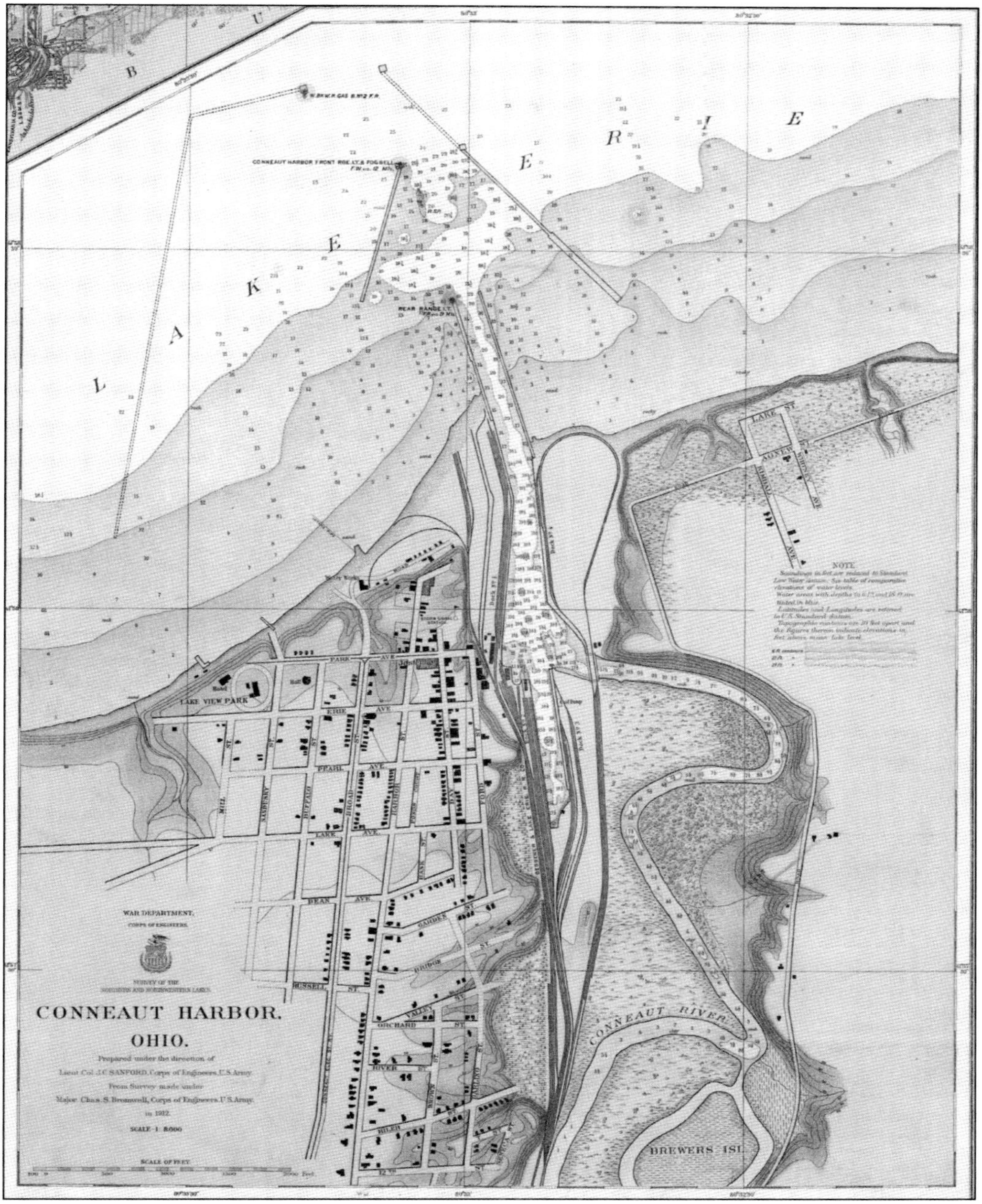

In this 1912 chart of Conneaut Harbor, the planned west breakwall, new lighthouse, and lakeward extension of the east breakwall are indicated as dotted lines. Also shown are the 1906 front and rear range lights, the Pittsburgh and Conneaut Dock Company docks, and Bessemer and Lake Erie railroad yards. The stone west breakwall would replace the existing rock-filled timber breakwall. In 1915, kerosene lights were replaced with acetylene gaslights.

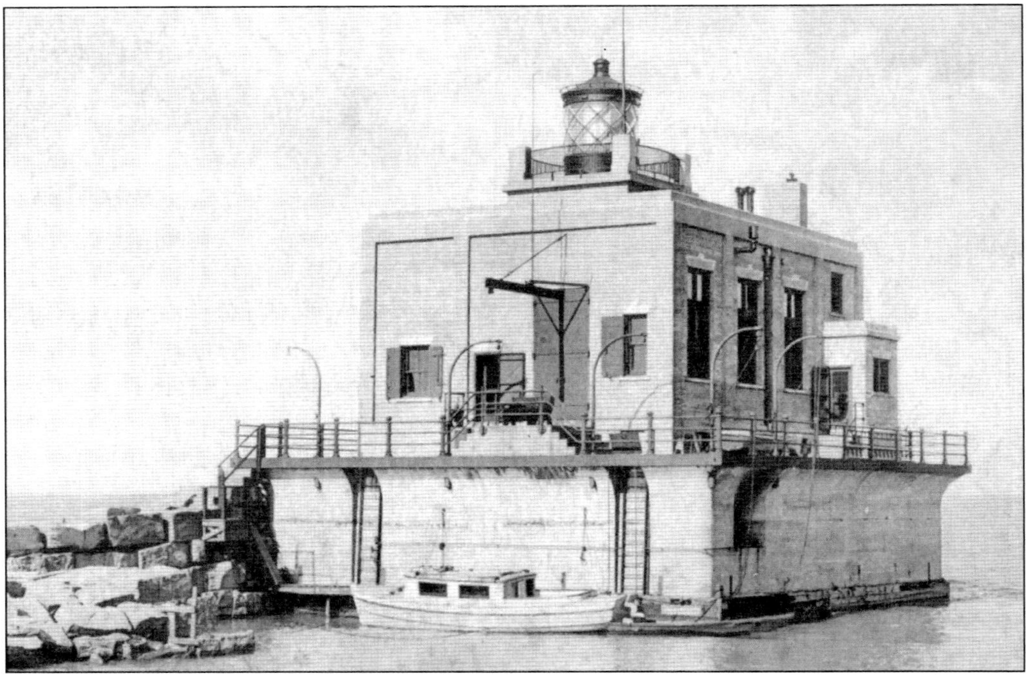

After years of delays due to World War I, the new light and fog signal station was illuminated for the first time on September 15, 1920. A 4,500-candlepower light, at 55 feet above the water, was visible for 15 miles. In 1921, Conneaut Harbor gaslights were converted to electric. By June 30, 1925, the final cost of construction was $89,666.67.

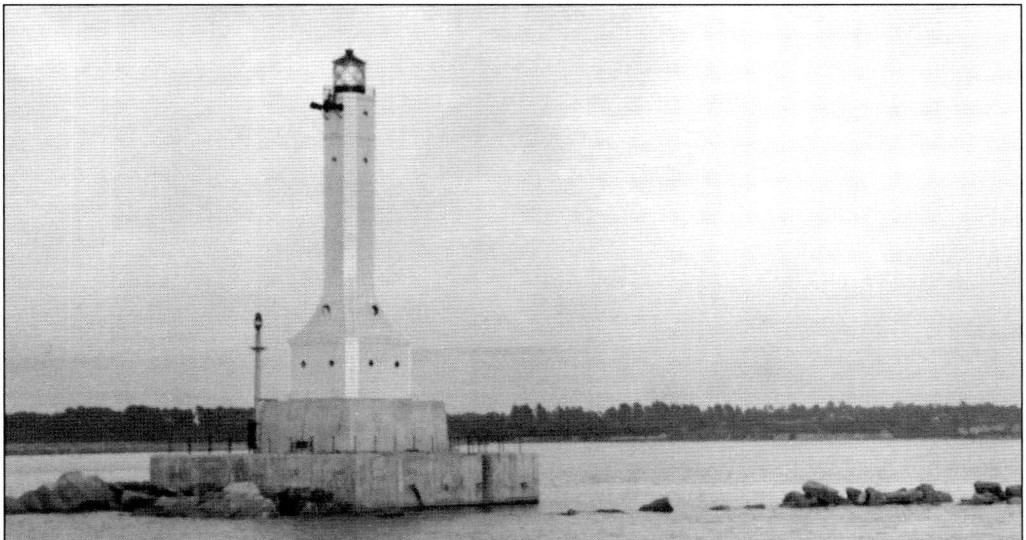

On April 1, 1936, the new Conneaut West Breakwater Lighthouse was established on the end of the west breakwall, where it stands today. The 1920 Conneaut Harbor Lighthouse was demolished using dynamite. The 925-foot lakeward extension of the east breakwall was completed the same year. At a cost of $70,000, the 1936 lighthouse stands 60 feet high and is made of white steel plates on a concrete foundation. (Historical Collections of the Great Lakes, Bowling Green State University.)

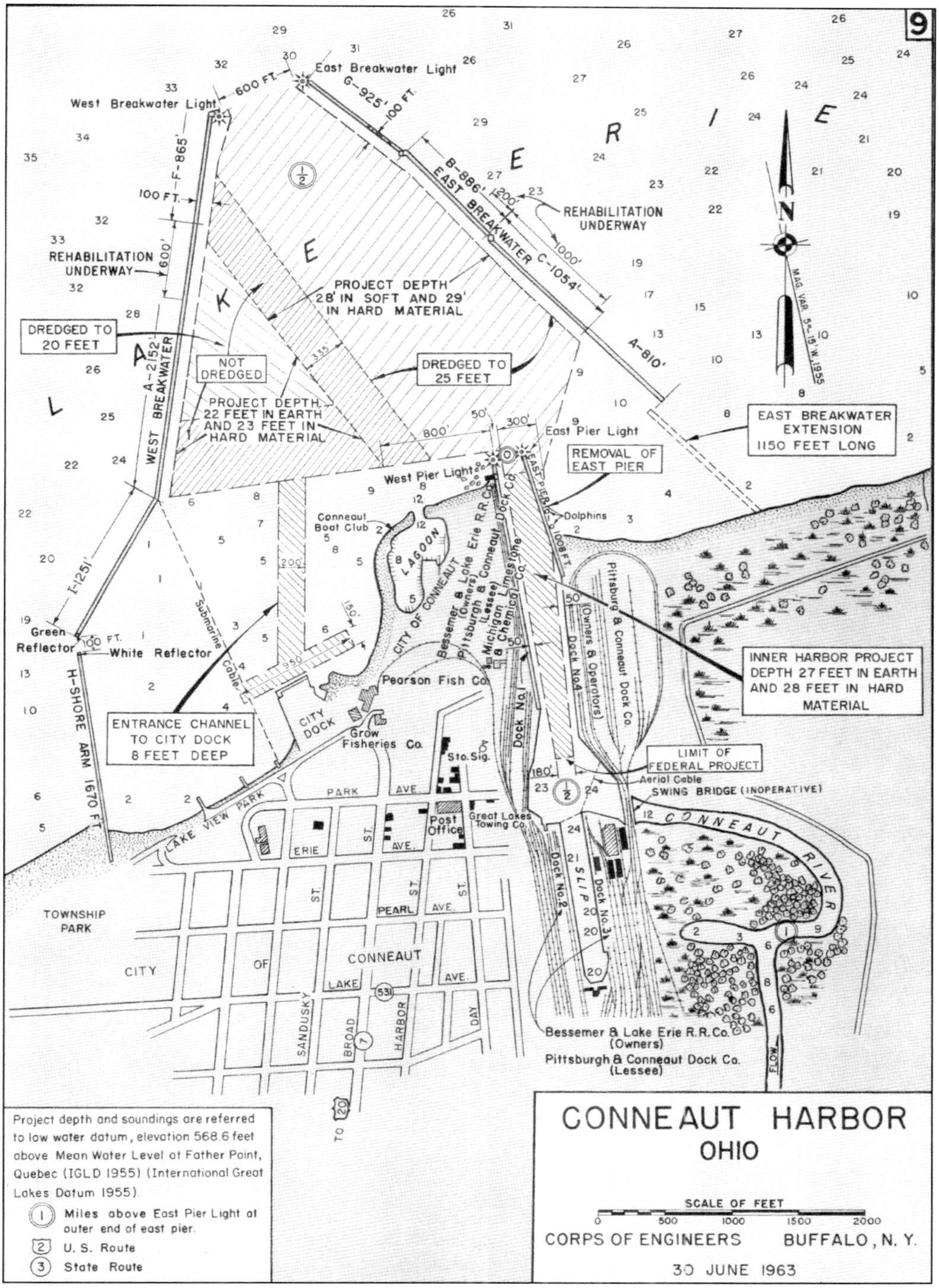

This 1963 harbor chart, in conjunction with the breakwall construction chart on page 13, details the chronology of the harbor construction. The chart also shows the location of docks one, two, three, and four, which will be of benefit when reading chapter two. The chart will also be beneficial when looking at the lakefront photograph on page 22. (U.S. Army Corps of Engineers.)

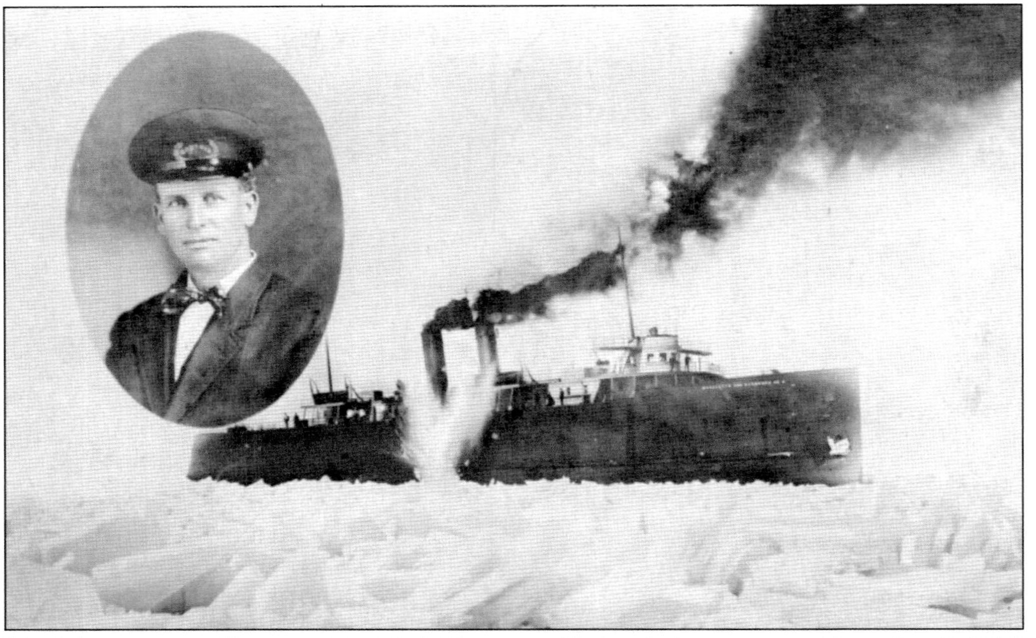

The *Marquette* and *Bessemer No. 2* left Conneaut Harbor before noon on December 7, 1909, bound for Port Stanley, Ontario, with a cargo of 30 or more railcars. Most of the crew members, including Capt. Robert McLeod (above, inset) were from Conneaut, Ohio, and the surrounding area. After leaving Conneaut Harbor, she sailed into a storm with winds in excess of 70 knots. The only passenger and the entire crew of 33 were lost. In 1910, the owners had an exact replica built and gave it the same name and identification number, 208021. Although the first ship had a single pilothouse, the second had a double-decker pilothouse.

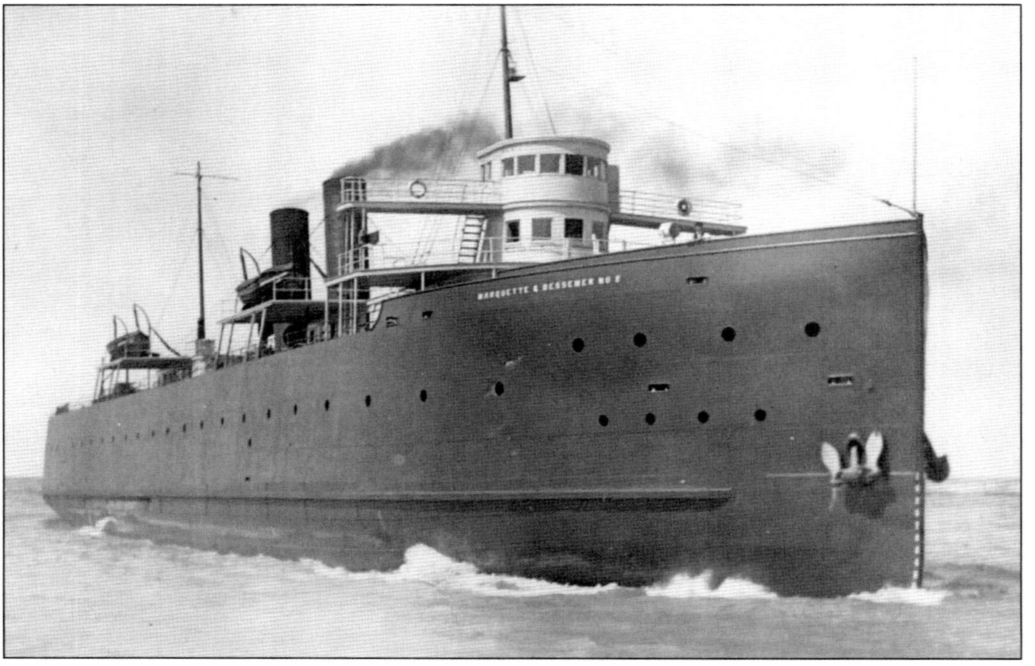

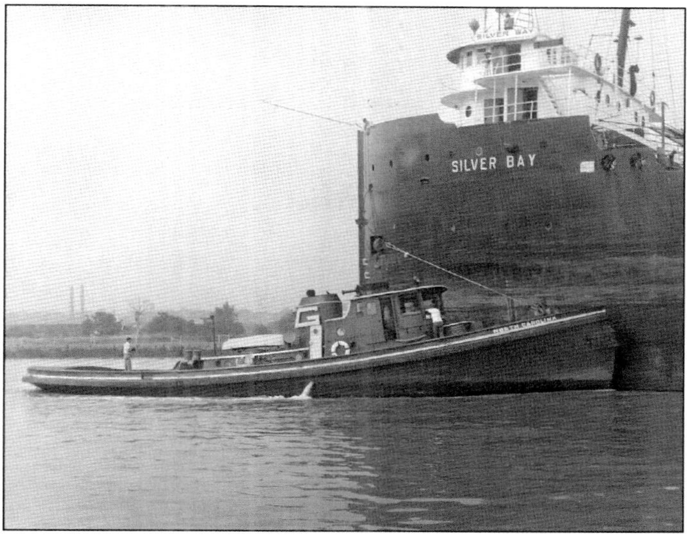

Leaving Cleveland on December 10, 1968, a three-man crew was homeward bound for Conneaut aboard the 81-foot-tug *North Carolina*. Before reaching Mentor-on-the-Lake, she began taking on water and the electrical system shorted. Without pumps or radio, the tug was sinking helplessly in the winter waves. Despite all attempts by her crew and the U.S. Coast Guard to save the tug, she went down in 32 feet of water. (Great Lakes Historical Society.)

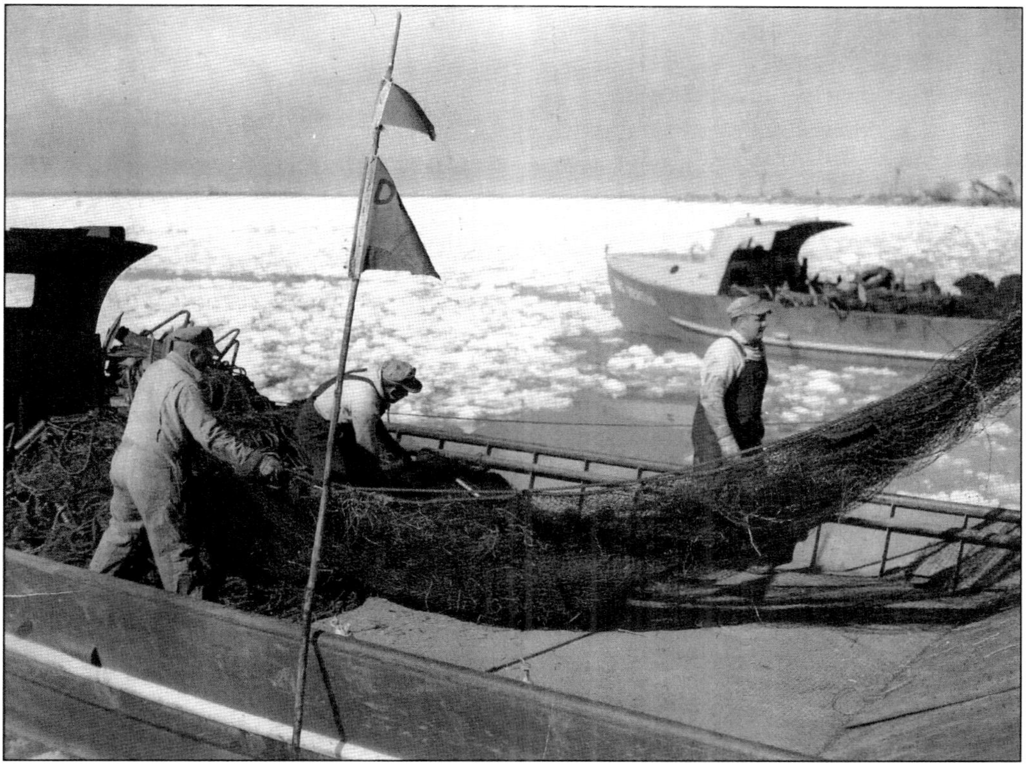

The commercial fishing season ran from March 15 through December 21. In the 1950s, there were about 21 trap-net fishing boats in Conneaut with each crew capable of catching 100 tons of fish per season. The fisheries accounted for approximately 100 jobs in Conneaut. In the 1980s, the U.S. government banned commercial fishing on Lake Erie to the glee of sport fisherman. Commercial fisheries on Lake Erie remain prosperous in Canada.

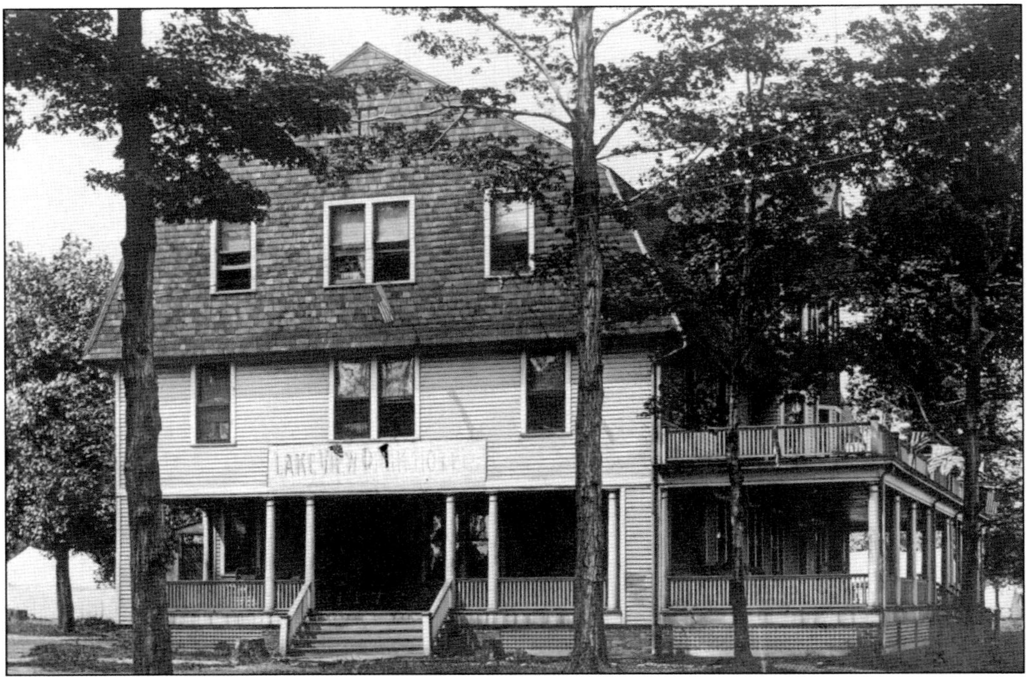

The Lake View Park Hotel was originally opened on June 23, 1899. It was purchased by the city and reopened on June 16, 1907. On March 4, 1937, a young boy, playing with matches, set fire to a pile of leaves, which quickly spread to the hotel, resulting in a total loss.

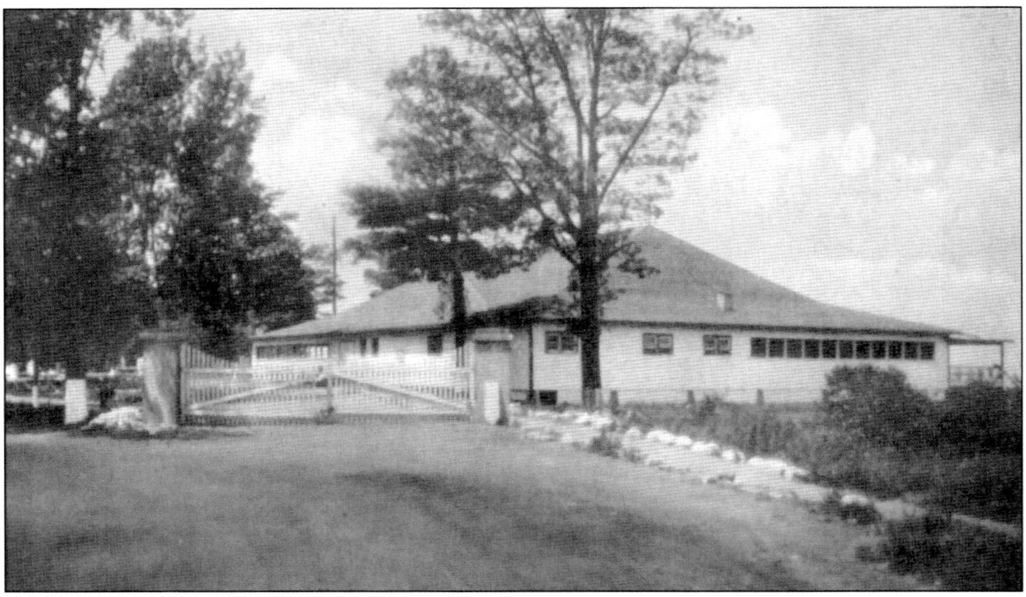

In this c. 1920 postcard, a dance pavilion with a veranda is seen, which once overlooked Lake Erie between the Lake View Park Hotel and the Sandusky Street entrance to the park. To the left of this entrance, there was a bowling alley, concession stands, and a roller-skating rink.

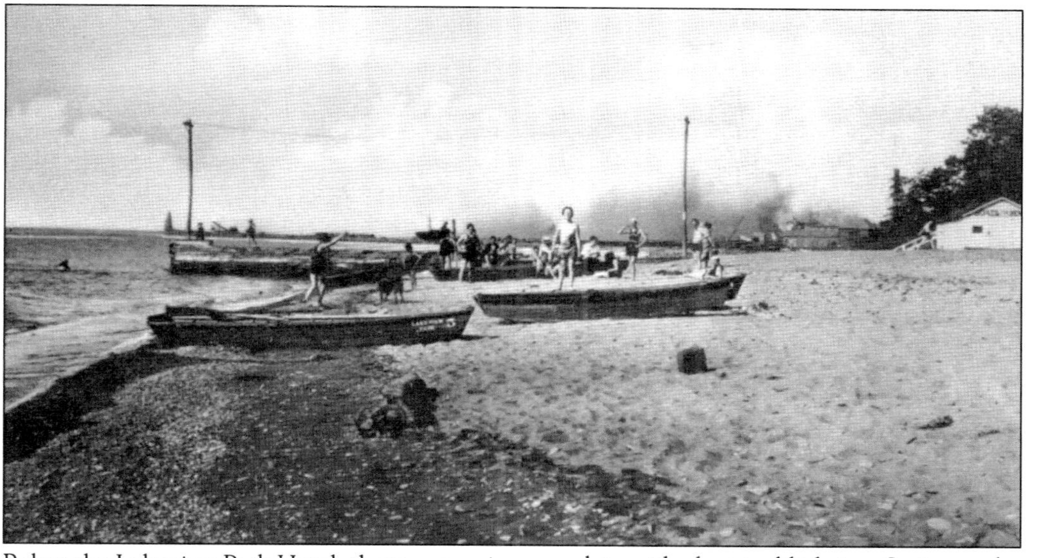

Below the Lakeview Park Hotel, there was a pier, complete with electrical lighting. Stairs, to the right in this postcard, lead up to center field of today's ball field. In addition to swimming at Lake View Park, boat rentals were also available. By the mid-1950s, the pier timbers had deteriorated, but a few concrete slabs still remained, and it became known as the "Old, Broken-up Pier."

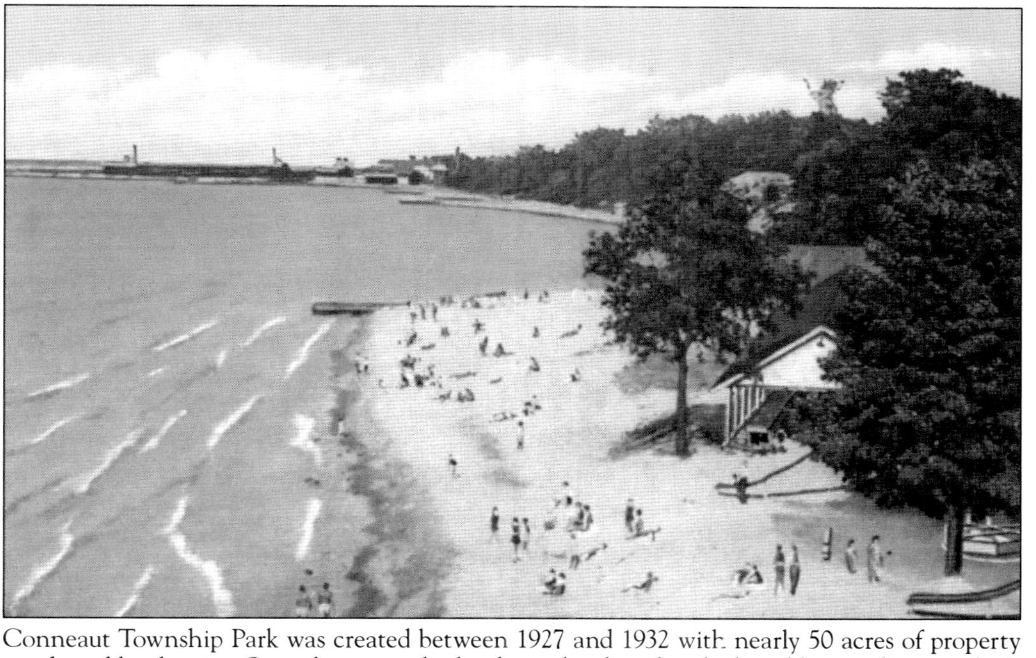

Conneaut Township Park was created between 1927 and 1932 with nearly 50 acres of property purchased by the city. Over the years, the land was developed with the addition of a caretaker's house, roads, storm and sewer lines, a bathhouse, a picnic pavilion, restrooms, a playground, tennis courts, a baseball field, and bleachers. In this early 1930s postcard, the west breakwall shore arm and public dock have not yet been built.

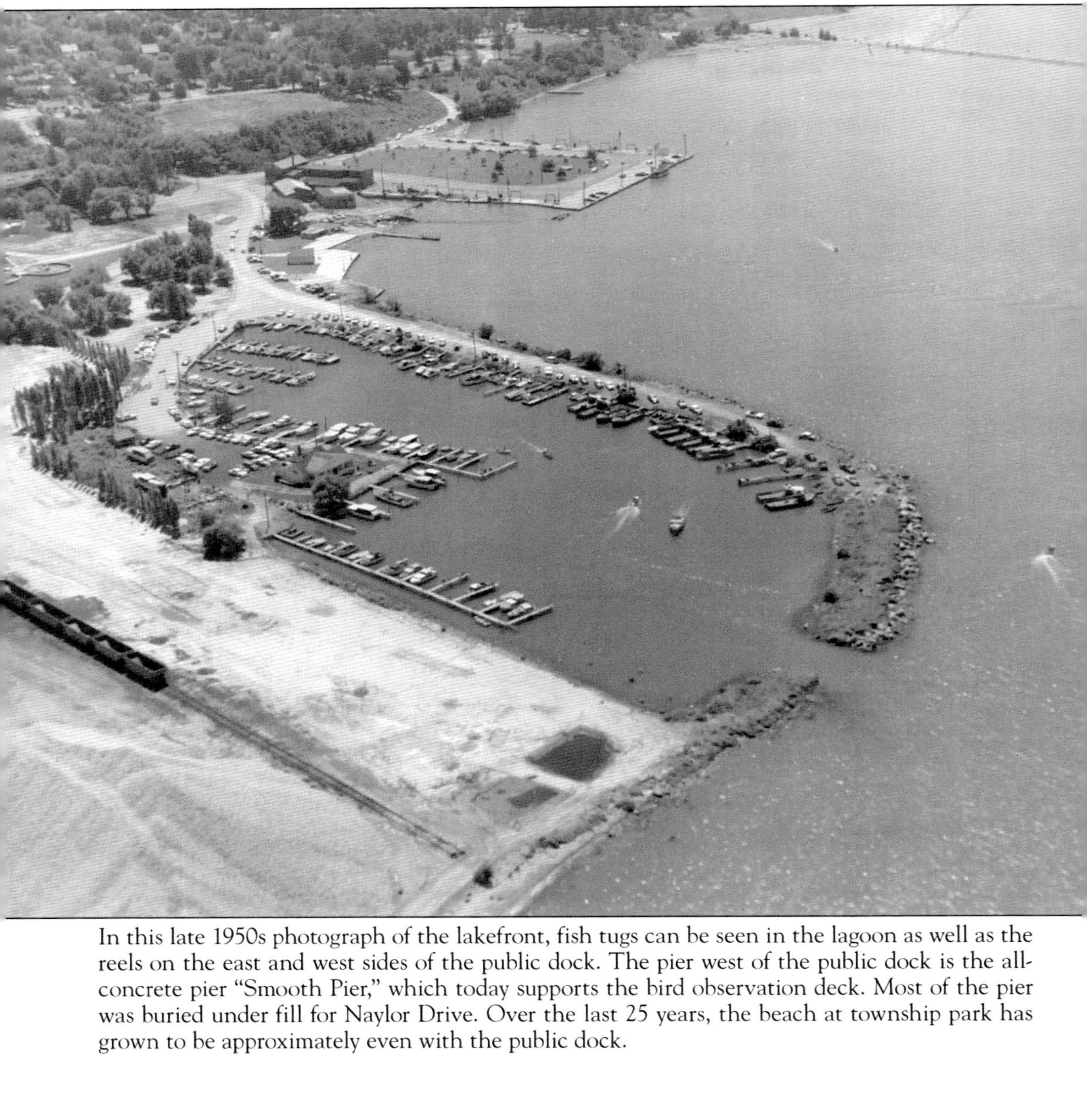

In this late 1950s photograph of the lakefront, fish tugs can be seen in the lagoon as well as the reels on the east and west sides of the public dock. The pier west of the public dock is the all-concrete pier "Smooth Pier," which today supports the bird observation deck. Most of the pier was buried under fill for Naylor Drive. Over the last 25 years, the beach at township park has grown to be approximately even with the public dock.

Two

The Docks

Due to railroad freight competition, Conneaut's importance as a Lake Erie port had vanished by the 1860s.

In February 1888, the Pittsburg, Shenango, and Lake Erie Railroad was in the process of extending the railroad to Lake Erie in order to supply iron ore from Minnesota to the steel mills in Pittsburgh. The citizens of Conneaut raised $25,000 to acquire 20 acres of land as an incentive for the railroad to build the terminus in Conneaut rather than Erie.

In the summer of 1892, the last spike was driven in the rails connecting Conneaut Harbor to the southwestern Pennsylvania steel mills. During the previous months, the railroad company had dredged a narrow channel and repaired the west pier. Upon the 600-foot west pier, six legs of Brown hoists, with 1-ton buckets, were erected along with two "Whirlies." Conneaut Harbor had been reborn and stood ready to unload iron ore from the Mesabe Range and ship it to the Pittsburgh steel mills.

In 1893, the railroad incorporated the Pittsburgh and Conneaut Dock Company as a subsidiary.

In 1897, Pittsburg, Shenango, and Lake Erie Railroad and other holdings were consolidated into the Pittsburgh, Bessemer, and Lake Erie Railroad under majority ownership of Andrew Carnegie.

In 1899, Andrew Carnegie decided to sell his steel interests and retire. Negotiations to sell for $157 million to Henry Clay Frick, and later to John D. Rockefeller for $250 million, failed to materialize. That same year, J. P. Morgan formed National Tube Company, controlling 85 percent of the country's steel tube and pipe production.

Carnegie responded by acquiring 5,000 acres of land in Conneaut and letting contracts for the construction of a $12 million steel tube plant capable of producing steel for $10 per ton less than National Tube.

In January 1901, Carnegie publicly announced his Conneaut plans. On February 26, Carnegie sent a handwritten letter to J. P. Morgan proposing a $480 million buyout. Morgan agreed and the U.S. Steel Corporation was incorporated on March 3.

In 1905, U.S. Steel completed an $8 million expansion of the former National Tube mill in Lorain rather than build a new tube mill in Conneaut.

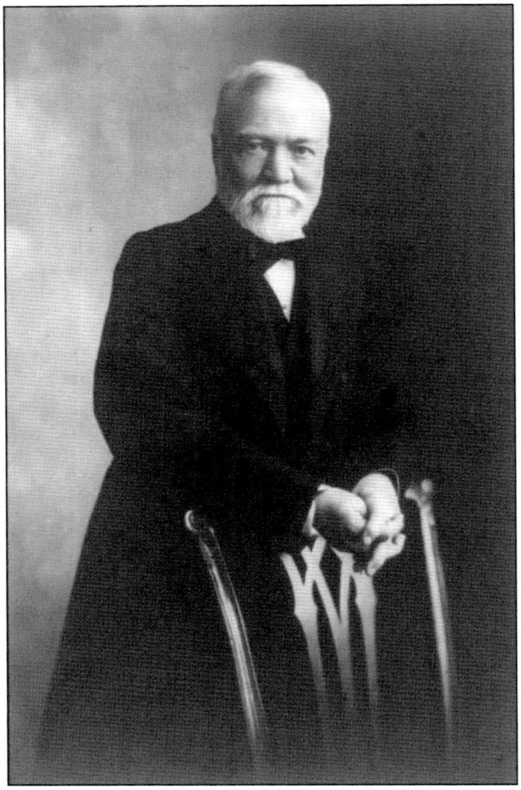

On November 25, 1835, Andrew Carnegie was born in Dunfermline, Scotland, to Margaret and William Carnegie, a weaver. Their home consisted of two rooms; William's handloom in one room and the family's living quarters in the other. By 1848, steam-powered looms had replaced handlooms, and the Carnegies immigrated to Pittsburgh. By 1901, Carnegie had become one of the wealthiest men in the world, second only to Cleveland's John D. Rockefeller. (Library of Congress.)

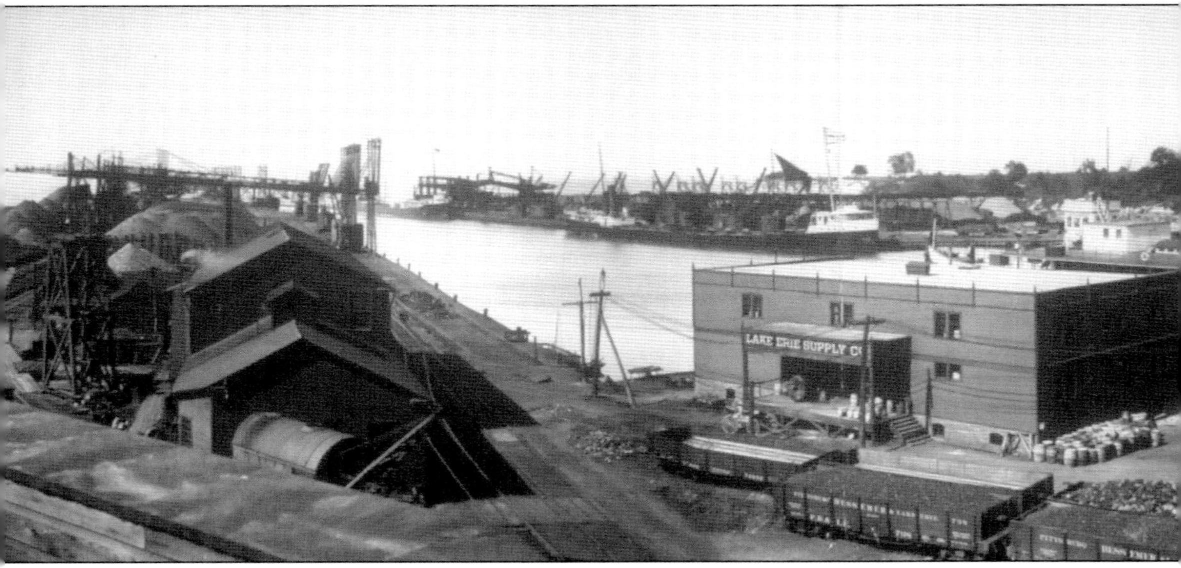

This 1901 panorama, by Conneaut photographers Naef and Joslin, represents the culmination of Carnegie's efforts to leverage Conneaut into a steel powerhouse. To the far left, beyond the ore piles, are Brown Hoists and King Hoists. Unloading the first ore carrier (left) are steam Huletts. Unloading the second are Brown Fast Plants. Under U.S. Steel, the Pittsburgh, Bessemer, and Lake Erie Railroad became the Bessemer and Lake Erie Railroad.

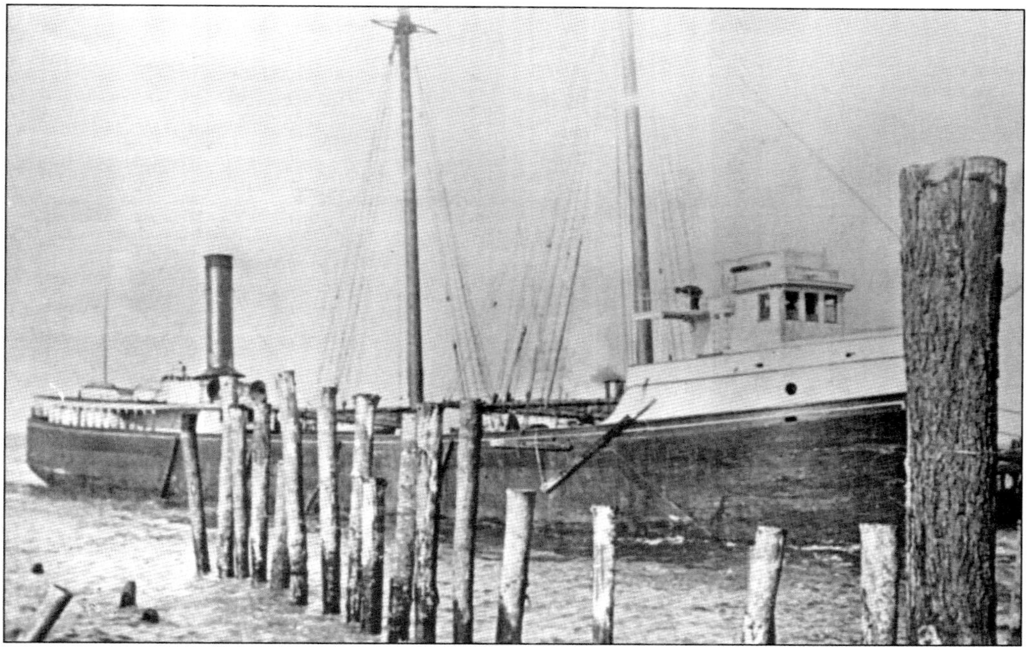

On November 3, 1892, the steamer *Charles J. Kershaw* arrived with 1,130 tons of iron ore from Minnesota. It was said that the entire population of Conneaut turned out to witness the event. When the ship was safely tied to the dock, every whistle on the docks sounded in celebration. The only other cargo unloaded that season was a deck load of pine lumber for the Record Manufacturing Company on September 30.

The two-story building houses the offices of the Pittsburgh, Bessemer, and Lake Erie Railroad and the Pittsburgh and Conneaut Dock Company. Behind the office building is a car carrier bound for Port Stanley Ontario. Coal gondola cars, waiting to be loaded, are lined up on a floating dock behind the car carrier. To the upper right is a coal dumper, which was used to load coal carriers bound for Duluth, Minnesota. (Library of Congress.)

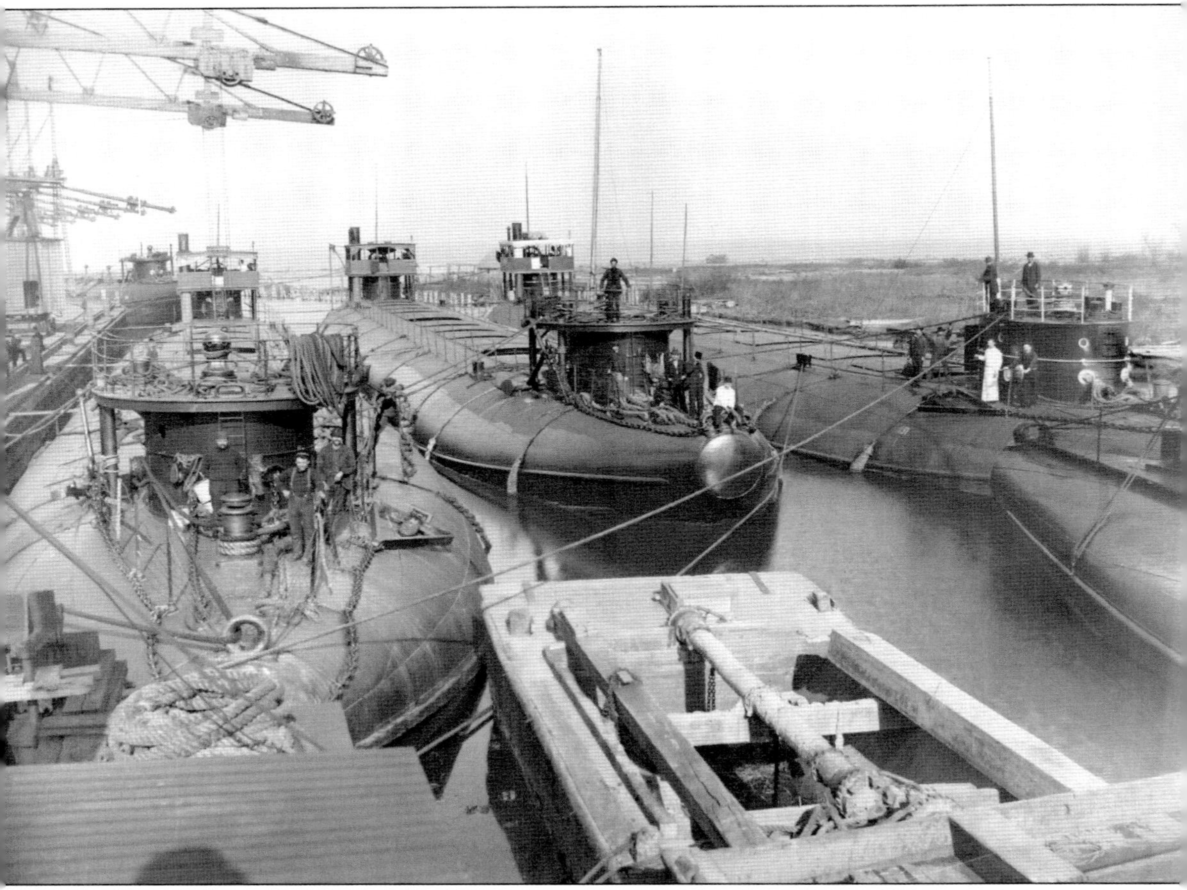

A total of 43 of these steamships, designed by Capt. Alexander McDougall, were built from 1887 to 1898. They were called "whalebacks" because only the rounded portion of the hull was seen above the waterline when fully laden. Since the bow bears some resemblance to a pig's snout, another derisive term it acquired was "pig boat." Due to their frequent presence in Conneaut Harbor, the area became known as "Pig Harbor."

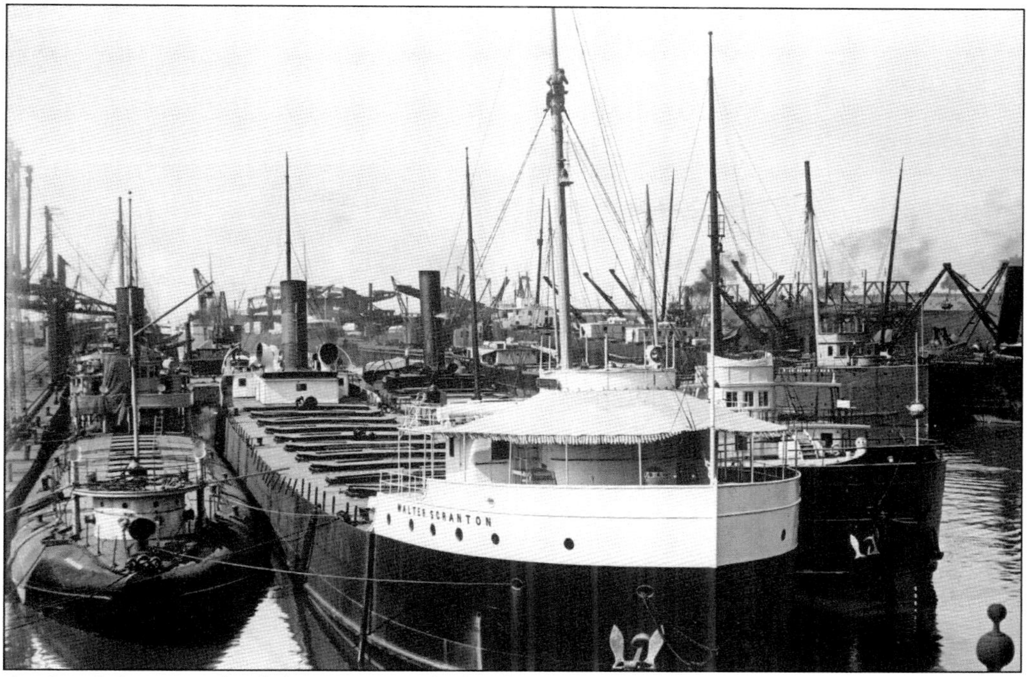

As the ability to unload ships improved, the ships grew larger. The larger vessels, such as the 435-foot *William Scranton* (pictured), could transport more tonnage down the lakes with the same size crew. However, the ships did not arrive at the docks at regularly scheduled intervals. It was not uncommon for ships to wait several hours before accessing a dock, resulting in the congestion shown here.

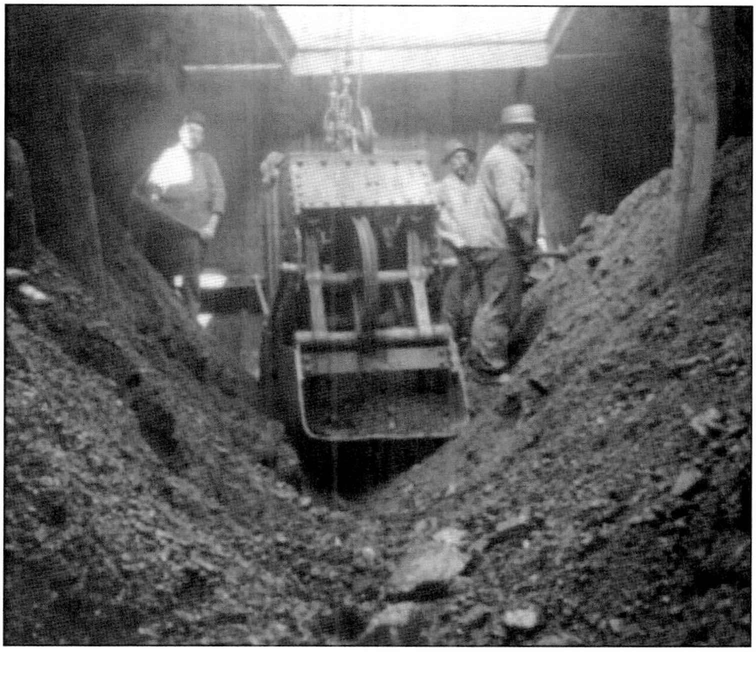

Both the King and Brown Hoists utilized a bucket, which was loaded into the hull of the ship and manually filled by men wielding picks and shovels. When the 1-ton bucket was full, the ore was raised out of the hold and transferred to a storage pile. In the summer, the heat inside the steel-hulled vessels and the shoveling of heavy iron ore created very unpleasant working conditions.

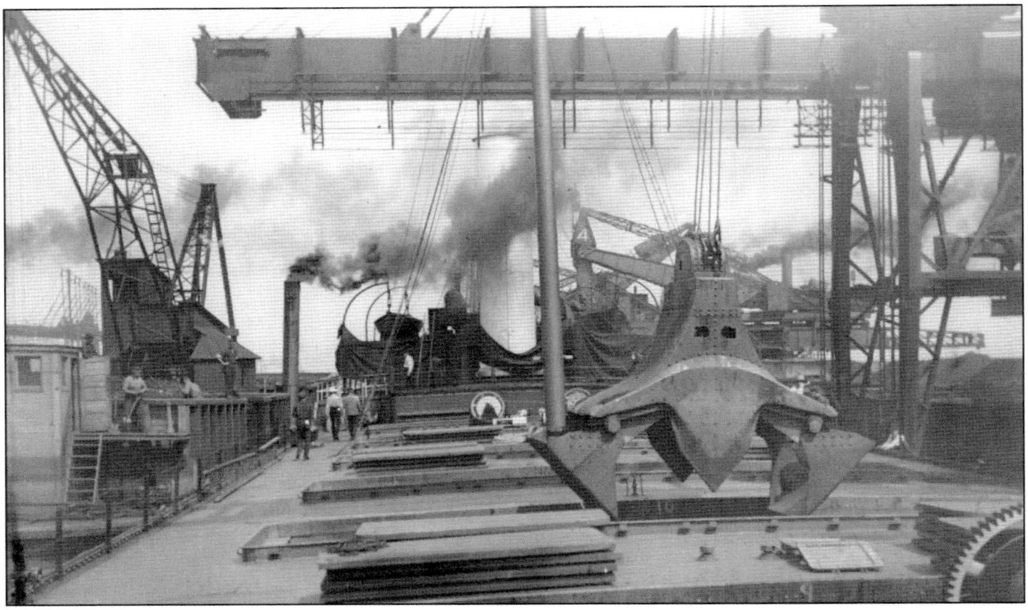

Another leap in productivity was achieved with the development of the Brown Fast Plants due to the elimination of manual bucket loading. The 5-ton bucket, with its shovel-like shape, would descend into the hold and bite into the iron ore with its jaws. The loaded bucket was then hoisted up and moved to the right, where its jaws would open and dump the ore into a storage pile.

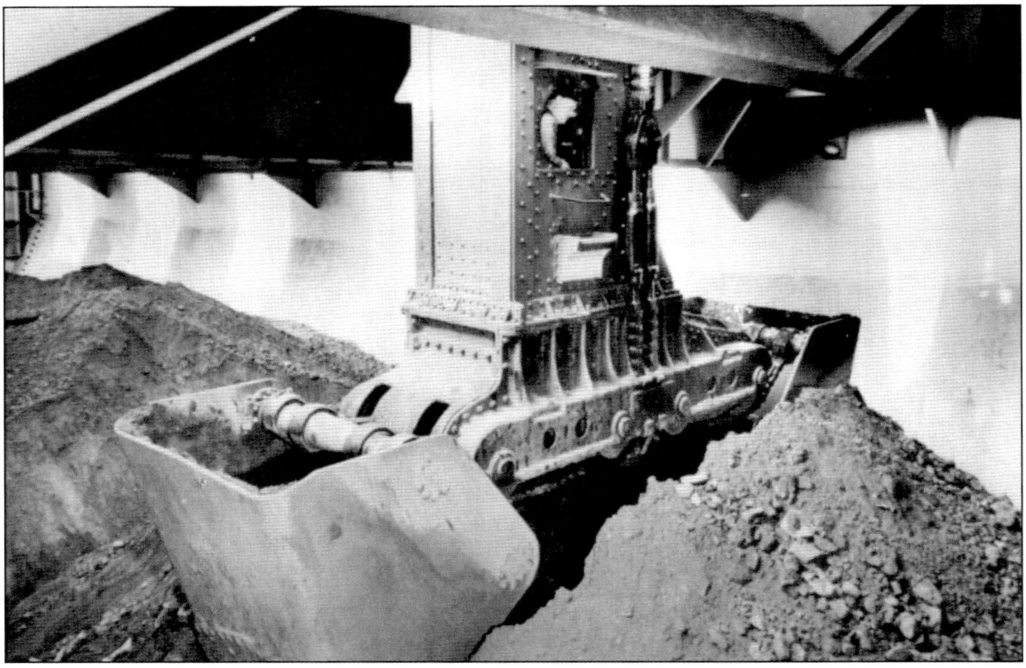

The Hulett iron ore unloader was patented by Conneaut-born George H. Hulett in 1898. Although Brown and McMyler Fast Plant operators could see into a ship's hold, the Hulett operators were able to descend into the hold atop the bucket. The ultimate advantage was the maneuverability, flexibility, and control the Huletts provided the operator. (Joyce Mackey Wright.)

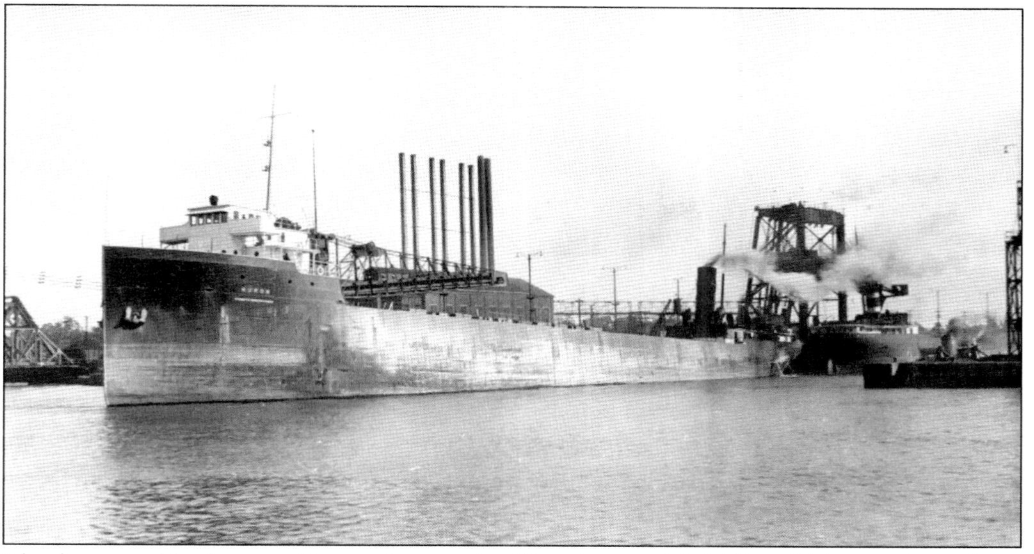

The first vessel built as a self-unloader was the *Wyandotte* in 1908. The 439.3-foot *Huron* (seen here) was built in 1914 and its twin, the *Conneaut*, was built in 1916. When the 1908 *Wyandotte* was scrapped in 1963, the *Conneaut* was renamed the *Wyandotte*. The building with the tall stacks, behind the *Huron*, is the Pittsburgh and Conneaut Dock Company's electric power plant.

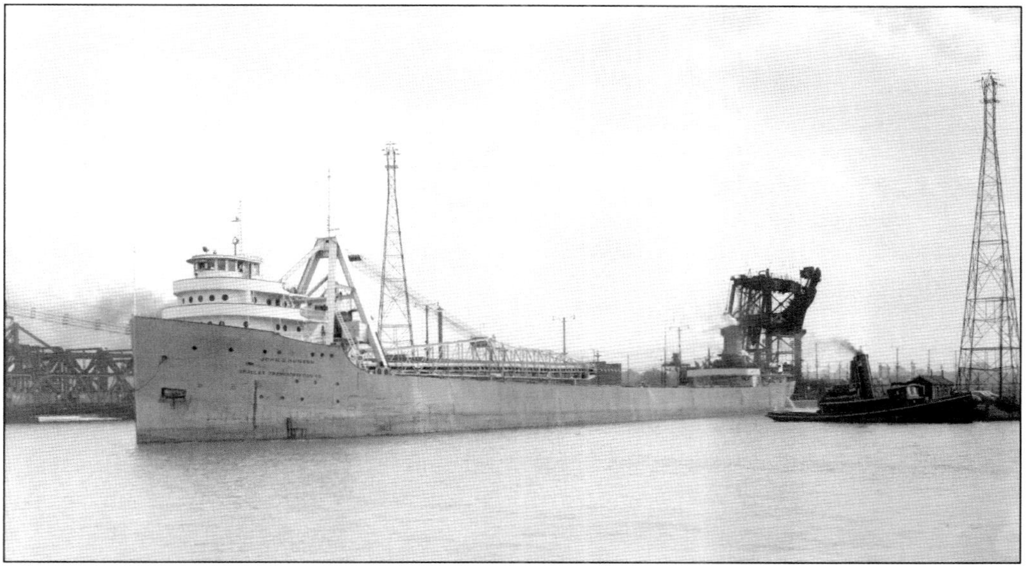

The 552-foot *Carl D. Bradley* was built as a self-unloader in 1917. She was renamed the *John G. Munson* in 1927. When electricity became available from Cleveland Electric Illuminating Company, it was no longer necessary for the Pittsburgh and Conneaut Dock Company to generate its own electricity, and the electric power plant was dismantled.

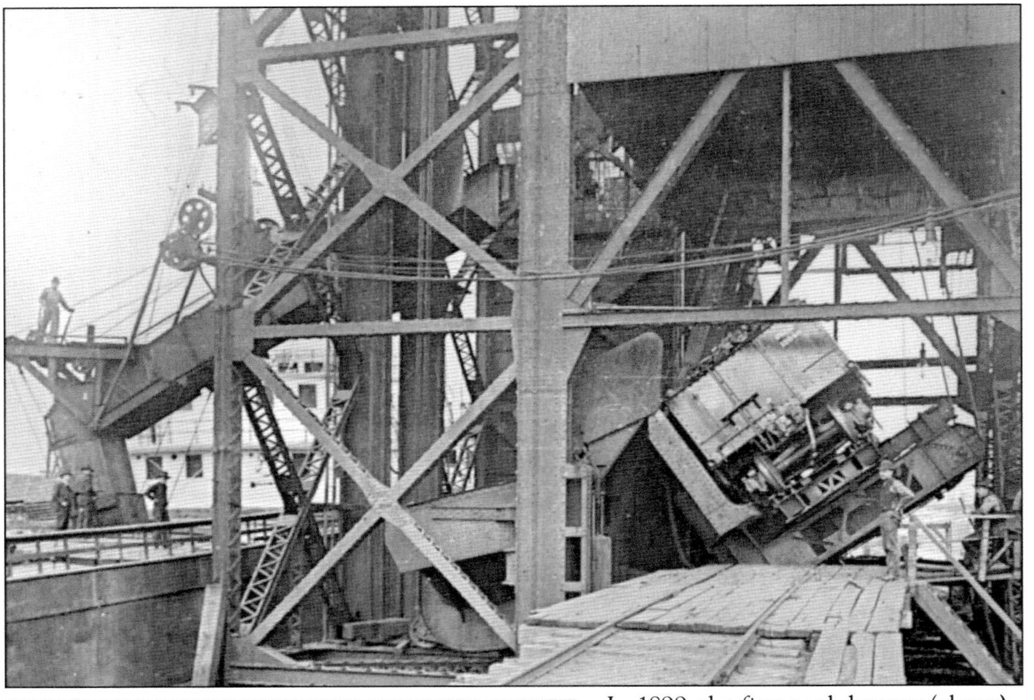

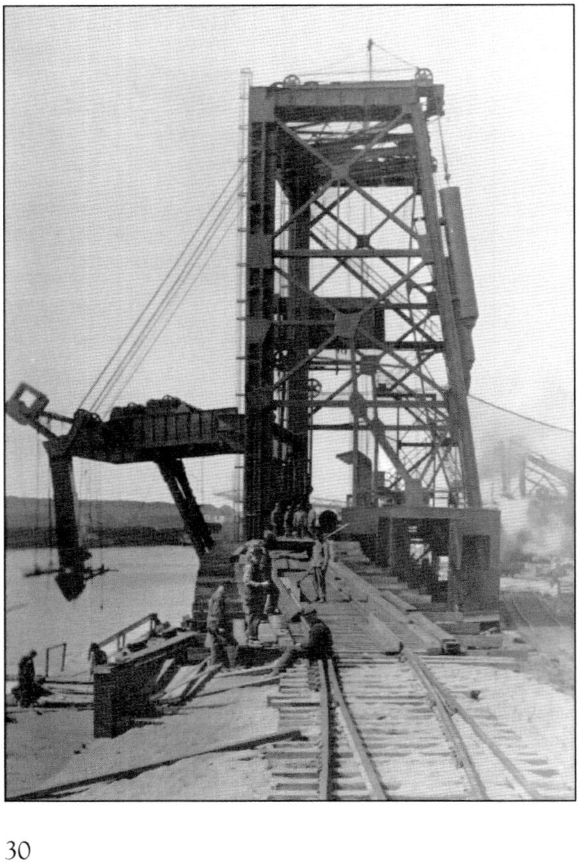

In 1899, the first coal dumper (above) was installed by the Pittsburgh and Conneaut Dock Company. It was erected on the east side of the slip, at dock three, and was capable of handling 25 cars per hour. In 1914, a new car dumper (left) was built at the same location. This McMyler coal dumper was designed by George H. Hulett and built in 1914. The 89-foot-tall structure was able to pick up and unload 100-ton railroad cars at the rate of one per minute. The boom operator had the ability to maneuver the chute and direct the coal. A four-man crew controlled the operation of the dumper and the movement of the railroad cars.

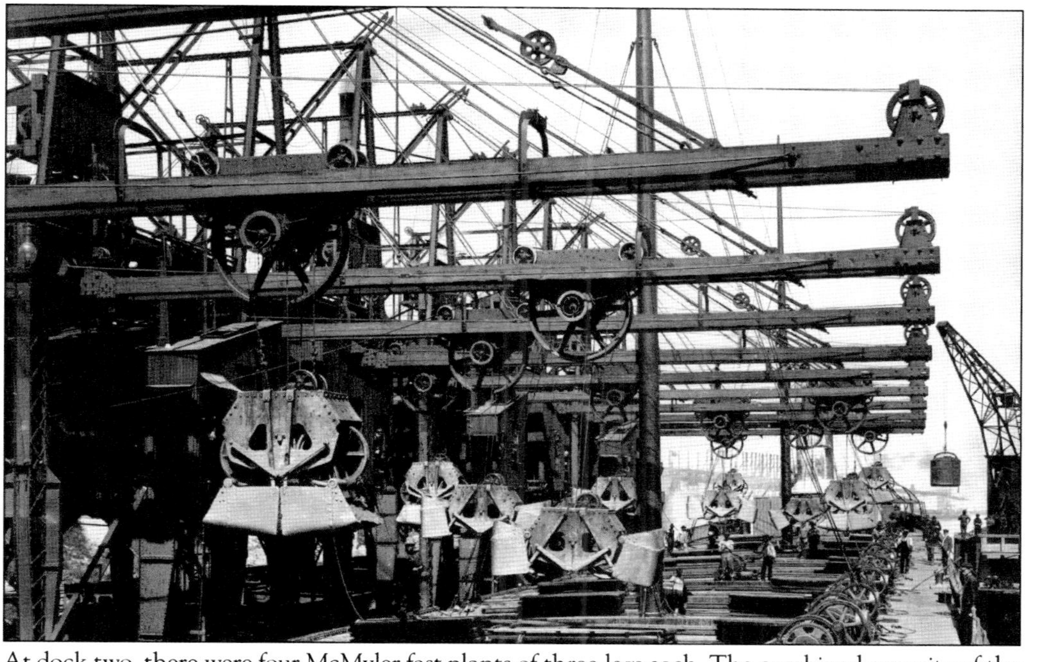

At dock two, there were four McMyler fast plants of three legs each. The combined capacity of the four fast plants was 400 tons per hour. In July 1903, these fast plants set a world record by unloading and forwarding 178,053 tons of iron ore in a single week. Using a steam locomotive and crew was an expensive method of moving railcars. This cost was eliminated through the use of a cable system. The cables can be seen in the photograph below between the men in the foreground and the railroad cars on the right. Power to the drum winch was provided by a stationary steam engine known as a mule, located in the building directly behind the McMyler fast plants.

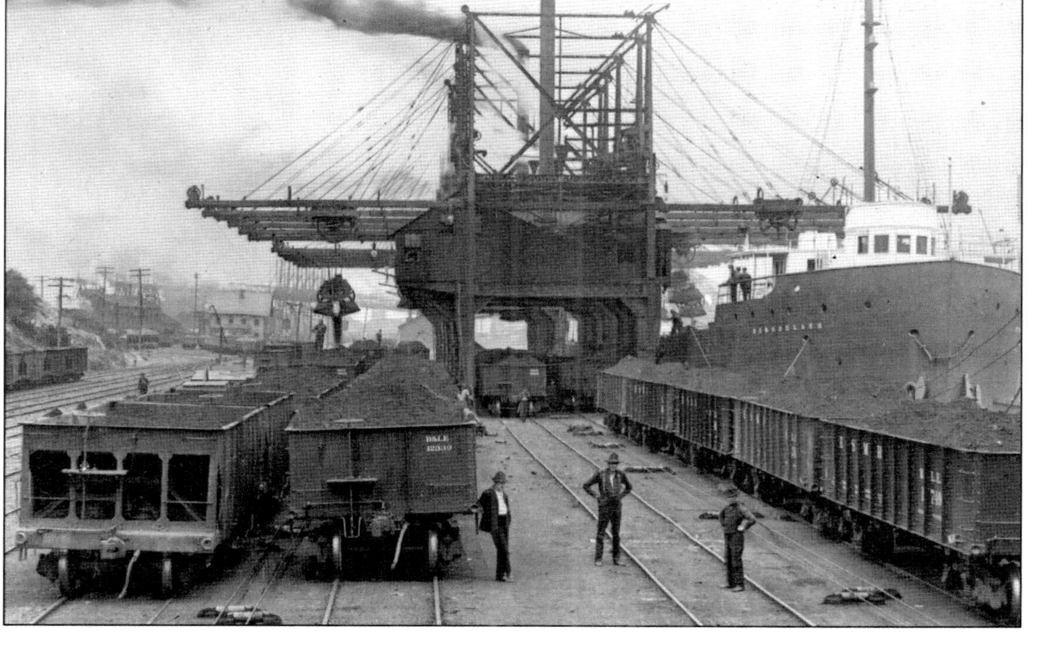

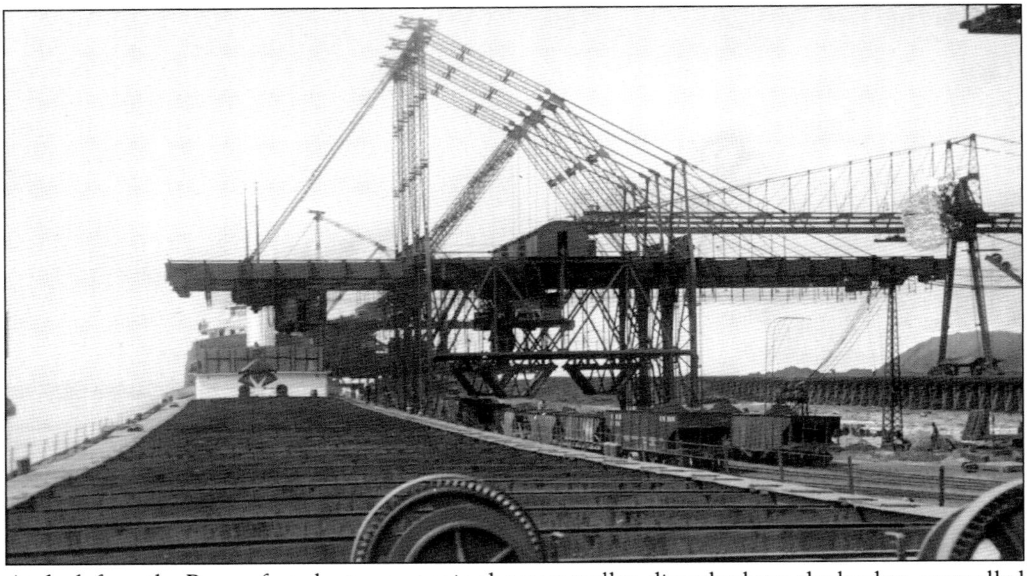

At dock four, the Brown fast plant operator in the man trolley, directly above the bucket, controlled all operations. The bucket was lowered into the hold with its scrapers (jaws) wide open. As the operator closed the scrapers, the bucket would fill. With its 5-ton buckets, Conneaut's four plants had a 7,200-ton daily unloading capacity. (Mary Jane Fritts.)

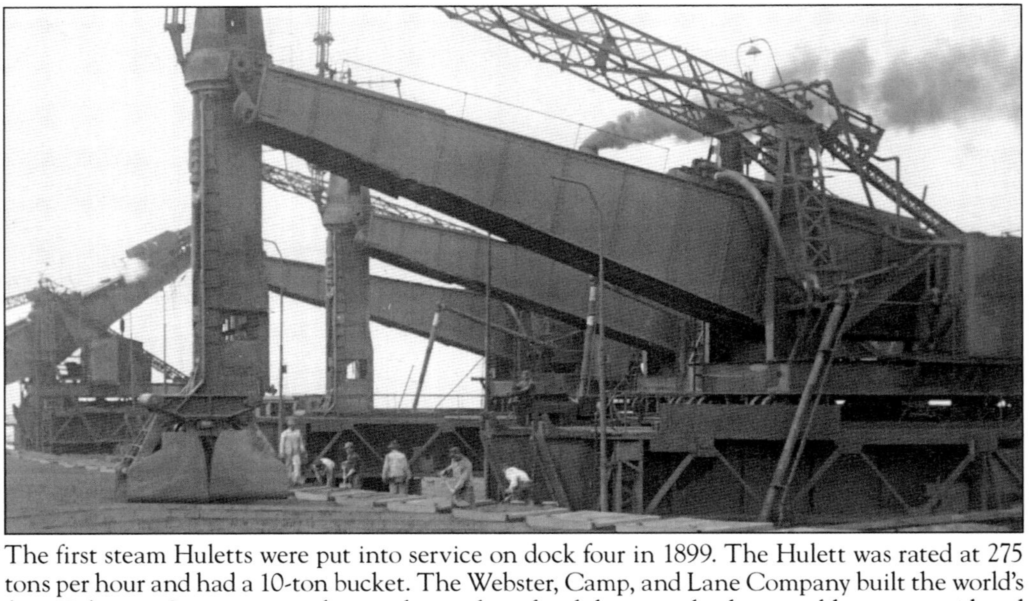

The first steam Huletts were put into service on dock four in 1899. The Hulett was rated at 275 tons per hour and had a 10-ton bucket. The Webster, Camp, and Lane Company built the world's first Hulett in Conneaut on the condition that if it did not work, they would not get paid and have to remove it at their own expense.

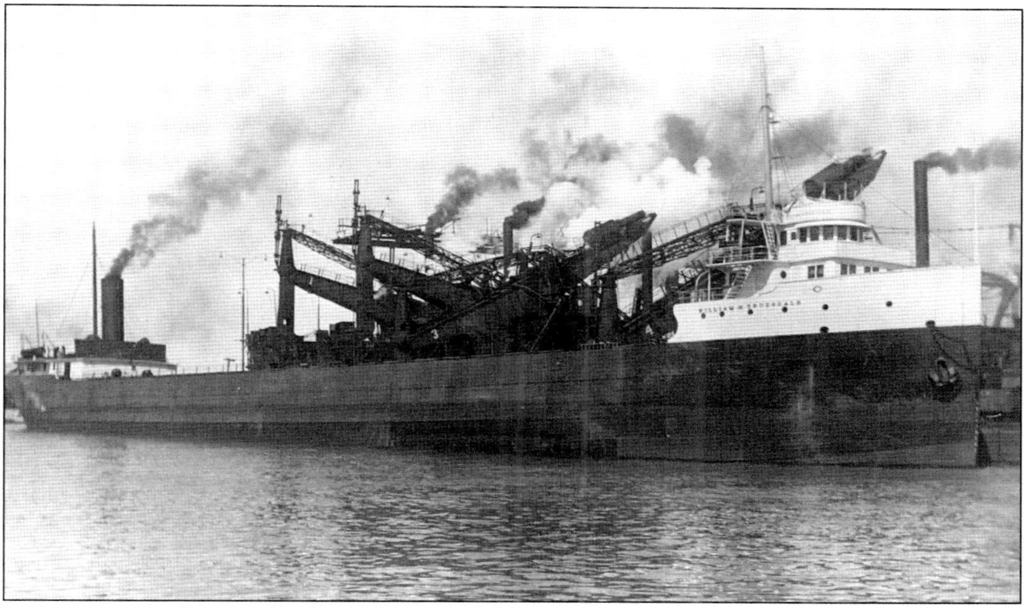

In this c. 1908 photograph, the 432-foot *William H. Truesdale* is being unloaded at dock four by four steam Huletts. The ship had a width of 52 feet, depth of 28 feet, and a carrying capacity of 3,764 tons. It would require nearly four hours to unload this vessel. The air pollution produced by the coal-burning steamships, Huletts, whirlies, locomotives, and homes of this era deposited soot on everything.

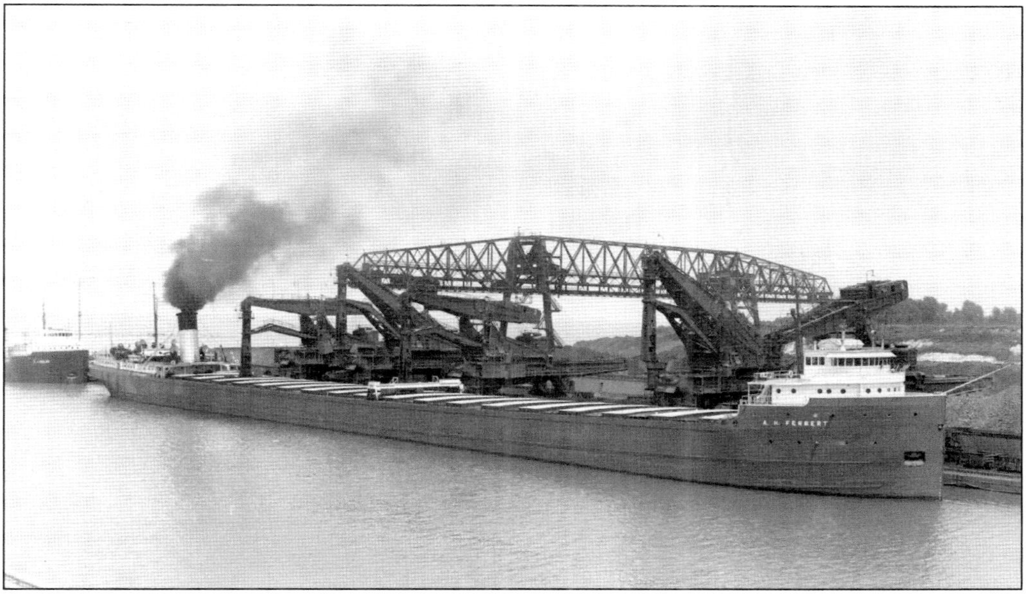

The first electric Hulett was installed at dock four in 1912. The electric Huletts, with the larger 17-ton bucket, could unload at a rate of 1,000 tons per hour. The 640-foot-long *A. H. Ferbert* (pictured) was built in 1942. It had a width of 67 feet, depth of 35 feet, and a carrying capacity of 17,700 tons. This U.S. Steel steamer could be unloaded in approximately four hours.

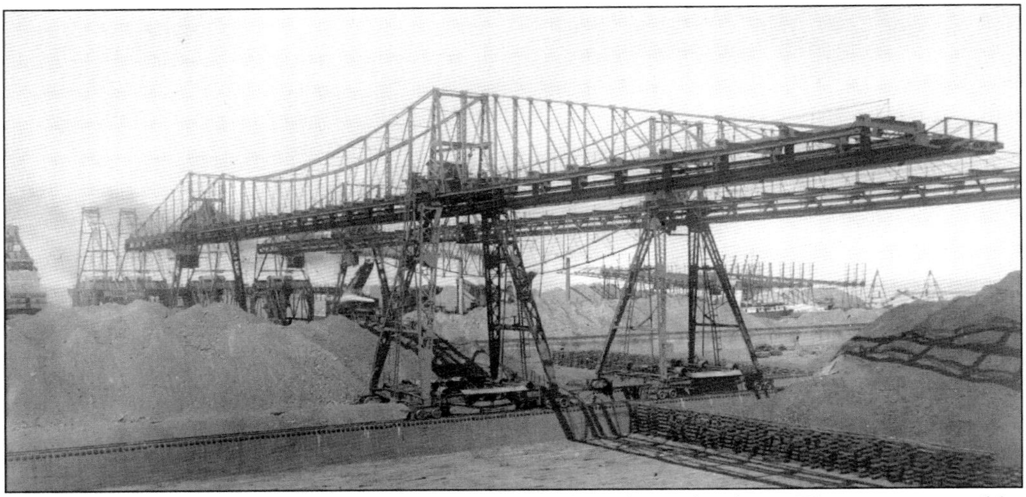

These iron ore bridges, at dock four, are 120 feet high and 600 feet long. They are operated by electricity with the cab and bucket similar to the Brown hoist. Their 7-ton buckets pick up ore from piles behind the Huletts and transfer it to storage piles. In the winter, the track-mounted bridge conveys ore from the storage piles into railroad cars for shipment to the Pittsburgh steel mills.

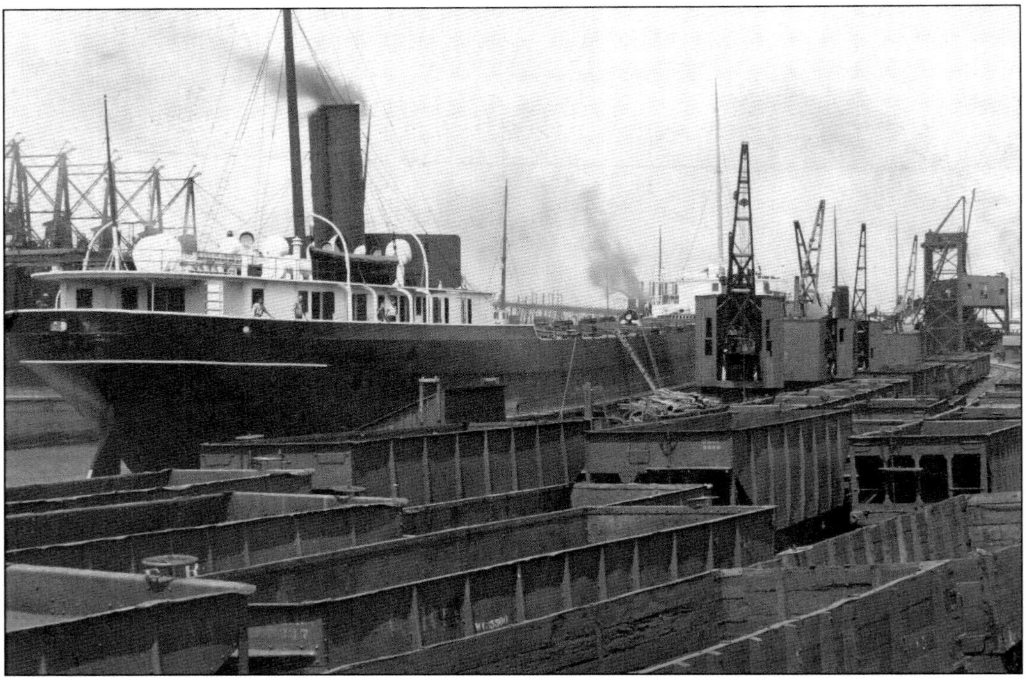

Whirly cranes (right), known as "whirlies," were self-propelled, steam-powered derricks with a movable cab and attached arm. They were wheel mounted and traveled along a track, which was wider than a railroad track, along the edge of the dock. Whirlies would be positioned alongside an open ship's hatch and load cargo into or unload cargo from a waiting railroad gondola. (Mary Jane Fritts.)

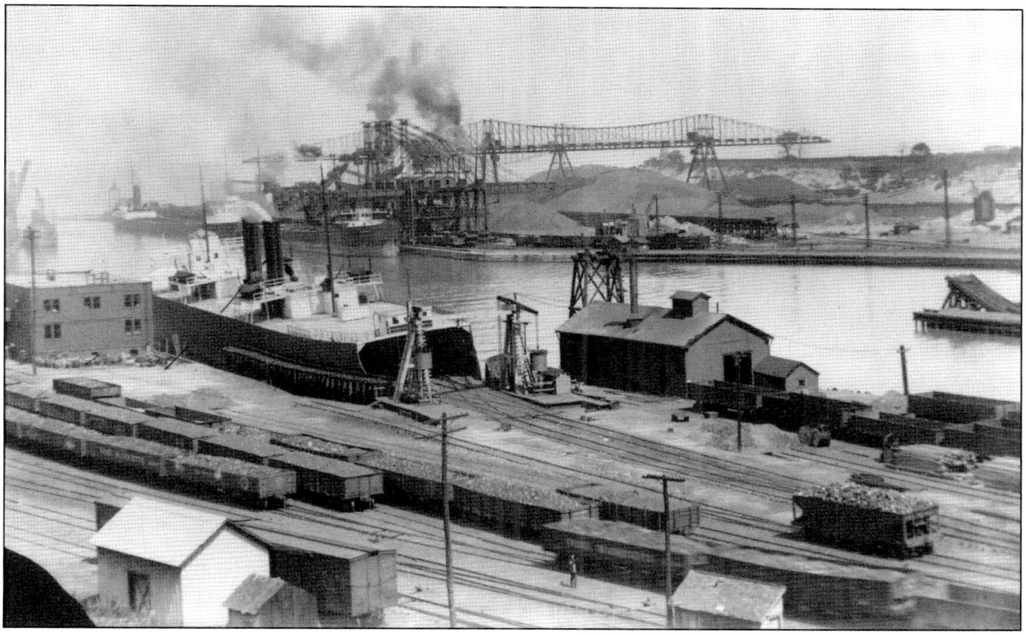

In this *c.* 1920 photograph, a car ferry is waiting to be loaded for its trip across Lake Erie to an Ontario port. On August 22, 1894, the first coal-laden Pittsburg, Shenango, and Lake Erie Railroad cars were ferried across Lake Erie to Port Dover, Ontario. Car ferries also ran between Conneaut and Port Stanley, Ontario. The car ferry service ended in August 1932.

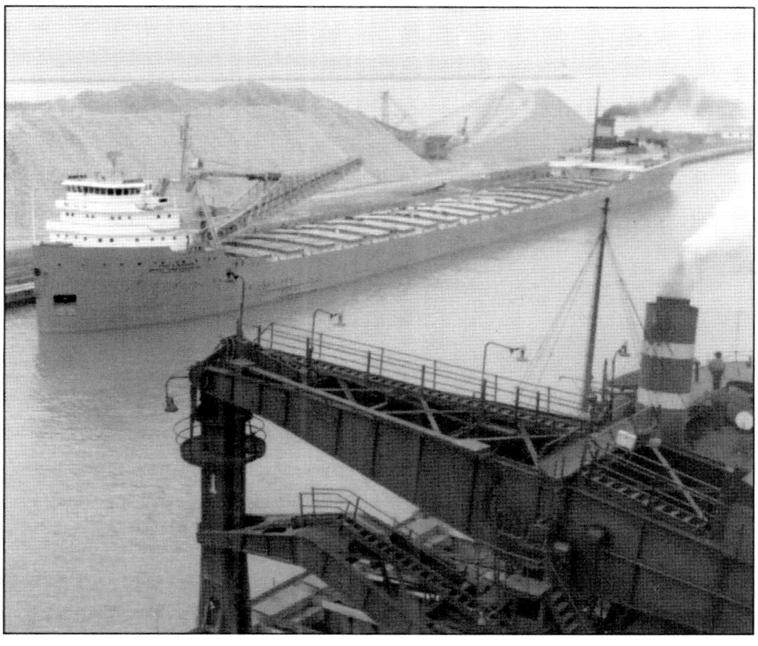

In 1946, the Michigan Limestone and Chemical Company took over dock one. Self-unloaders were relegated to coal and stone during the first half of the 20th century. By the 1970s, almost all ships were equipped with their own unloading gear. The last straight deck freighter was launched in 1985. This 1960s view of dock one from atop the Huletts on dock four signifies the end of an era.

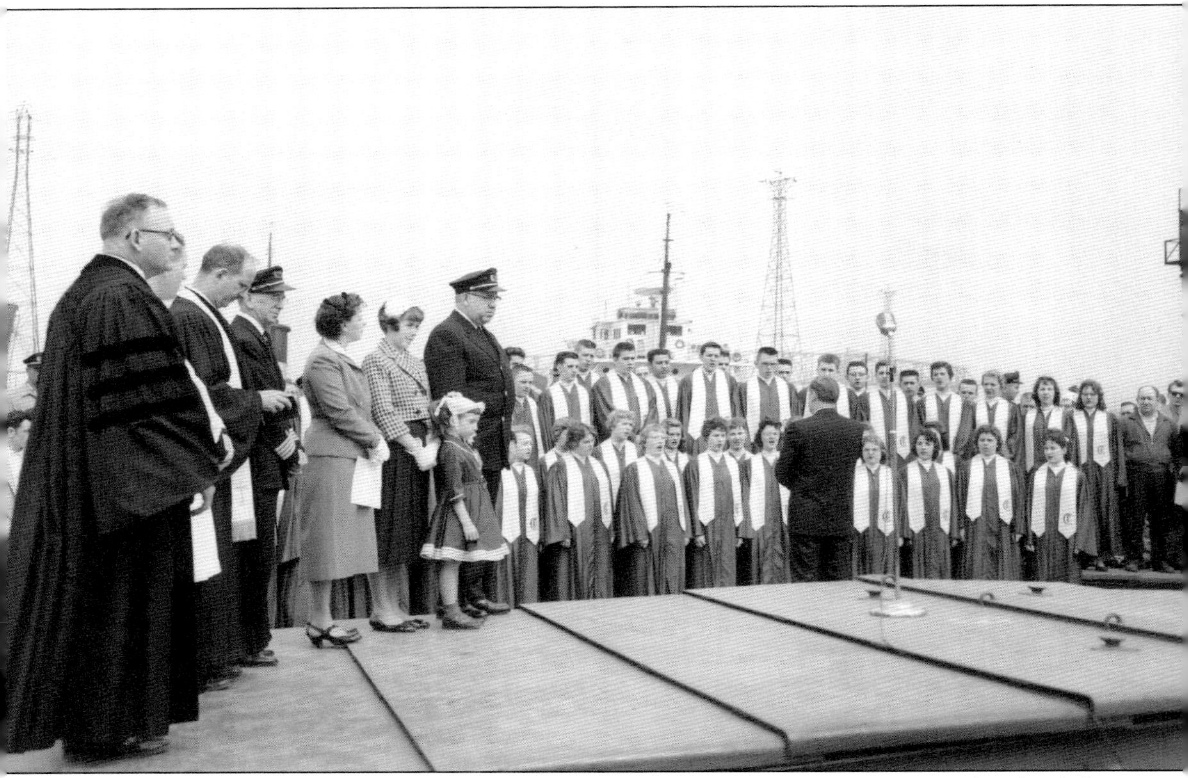

In this *c.* 1959 photograph, local clergyman, sailors' families, and the Conneaut High School choir participate in the annual Blessing of the Fleet. This tradition, which took place each spring, was meant to ensure a safe and prosperous shipping season. The choir is being directed by Kaarlo Mackey, who was the choir and band director from 1935 until 1961 except for four years with the U.S. Army Band during World War II.

Three

Industry

Most of the early industries in Conneaut were small operations employing a handful of men. These enterprises consisted mainly of sawmills and gristmills, which harnessed waterpower by building dams on Conneaut Creek.

The largest early industry was Ohio Furnace, which was founded by A. Dart and M. P. Ormsby in 1833. Ohio Furnace employed about 150 men and was located on Conneaut Creek near the Furnace Road bridge. The company produced stoves and other cast-iron utensils from bog ore, which was gathered locally from an area known as the Lake Plain. The most likely reason for the location on Conneaut Creek was to power a sawmill, which was used for the production of the charcoal needed to melt the bog ore. Ohio Furnace operated at this location until 1855.

The Conneaut River Paper Mills was located at the foot of the East Main Street hill on the west side of Conneaut Creek. A dam was built approximately halfway between Main Street and the U.S. Route 20 viaduct to provide power for the mill. From the time the mill opened in 1855, it was an unprofitable venture. In 1872, Sagito J. Smith purchased the mill and installed new equipment. By manufacturing and printing flour sacks, he was able to sell direct to millers and the business became profitable. On December 11, 1892, the paper mill was completely destroyed by fire.

In the 1860s, with the harbor in decline, community leaders became committed to moving Conneaut forward. As Conneaut grew, it became more attractive to other businesses, igniting a positive and self-reinforcing cycle of continued growth.

In 1882, community leaders were able to convince the New York, Chicago and Saint Louis (Nickel Plate) Railroad to build its division headquarters in Conneaut. With intense competition from other communities, Conneaut was selected when 70 acres of land was given to the Nickel Plate to build their facilities.

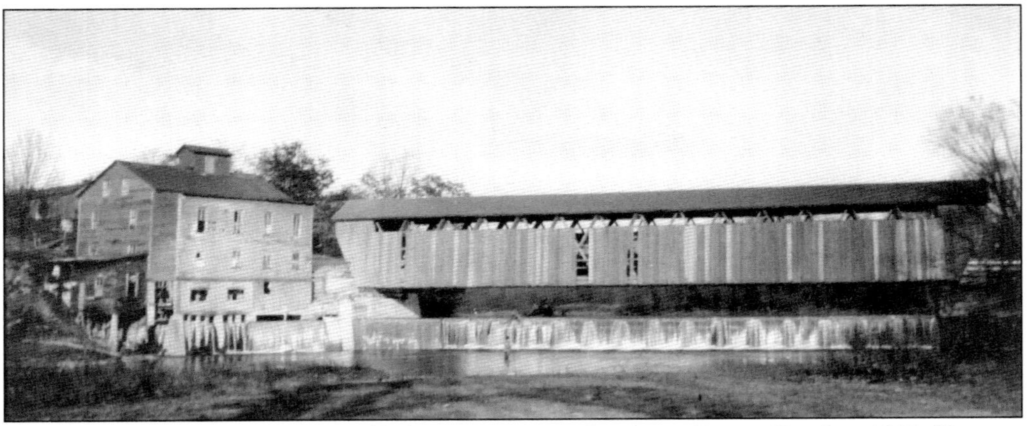

Elisha Farnham built this gristmill along Conneaut Creek on Center Road in 1841. His son, Patrick Henry Farnham, expanded the operation and built two new dams in 1878. This covered bridge was replaced by a concrete arch bridge in 1929. Its sister bridge and dam, to the south, were replaced with rubble and fill. The mill was destroyed by fire in the 1930s.

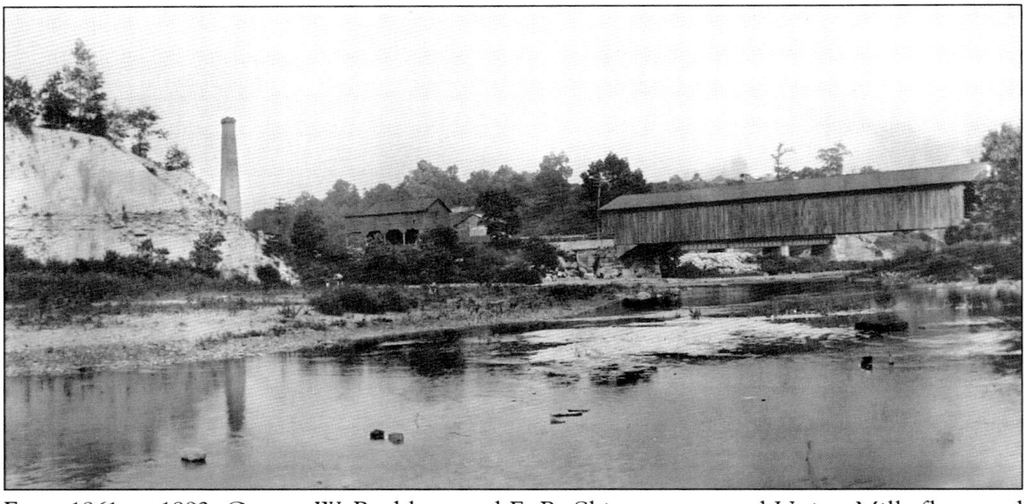

From 1861 to 1892, George W. Rathbun and E. R. Skinner operated Union Mills flour and gristmill, where the chimney stood at the foot of the Mill Street hill. C. H. Best acquired the mill and operated it as a feed and cider mill before moving his business to the east side of Mill Street. The 154-foot-long covered bridge (right) was built around 1867 and was replaced with a concrete arch bridge in 1925.

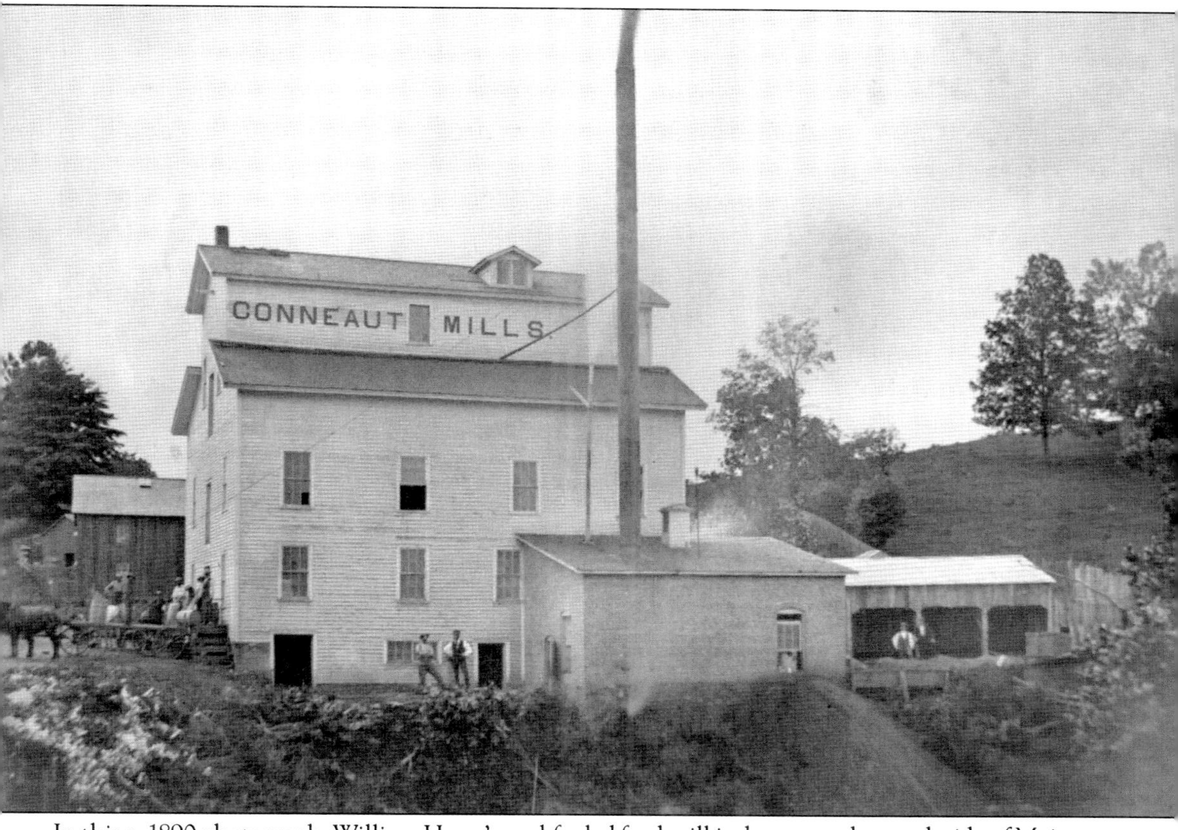

In this *c.* 1890 photograph, William Heyer's coal-fueled feed mill is shown on the north side of Main Street at the bottom of the hill. Conneaut Paper Mill, located south of Main Street, was destroyed by fire on December 11, 1889. A dam, which existed north of today's viaduct, provided the paper mill with waterpower. The pier of the Main Street bridge can be seen under the horses.

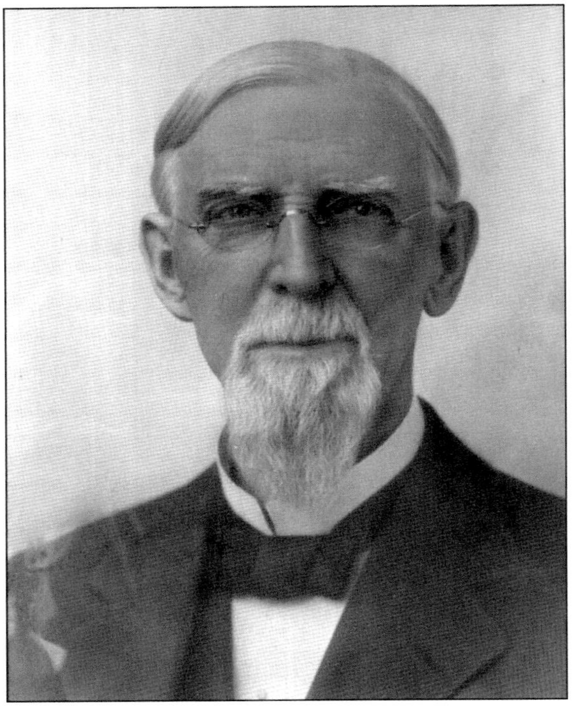

George John Record was born on August 22, 1839, raised on the family farm in Hanover, New York, and educated in common schools and Fredonia College. He and his wife moved to Conneaut in 1867, where he established a hardware store. He designed an improved method of packaging and began with one man making tin-lined butter packages in his hardware store. By the time he sold his company, he held 28 patent rights on his own inventions.

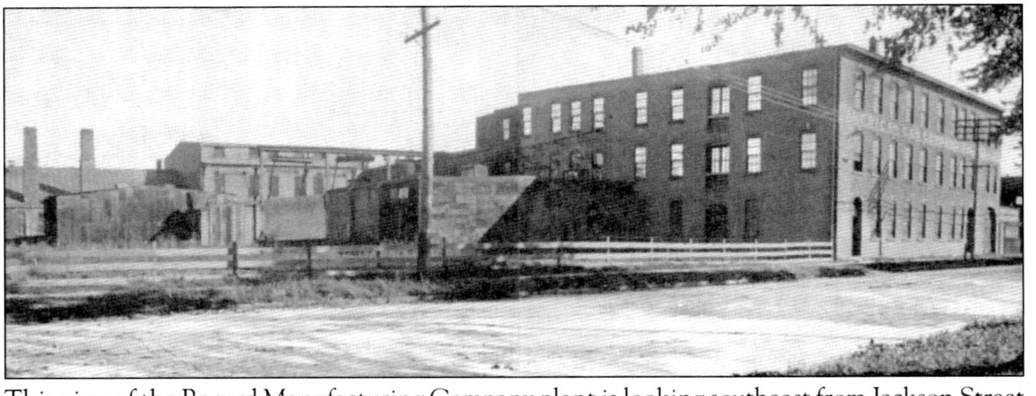

This view of the Record Manufacturing Company plant is looking southeast from Jackson Street toward the office at 362 Broad Street. Among the articles manufactured in the factory were tin-lined butter packages, shipping cans, sap spouts, pails, and sugar-maker's supplies. George Record opened his factory with 12 employees in 1882. By the time he sold the company to the American Can Company in 1901, the number of employees had grown to 200.

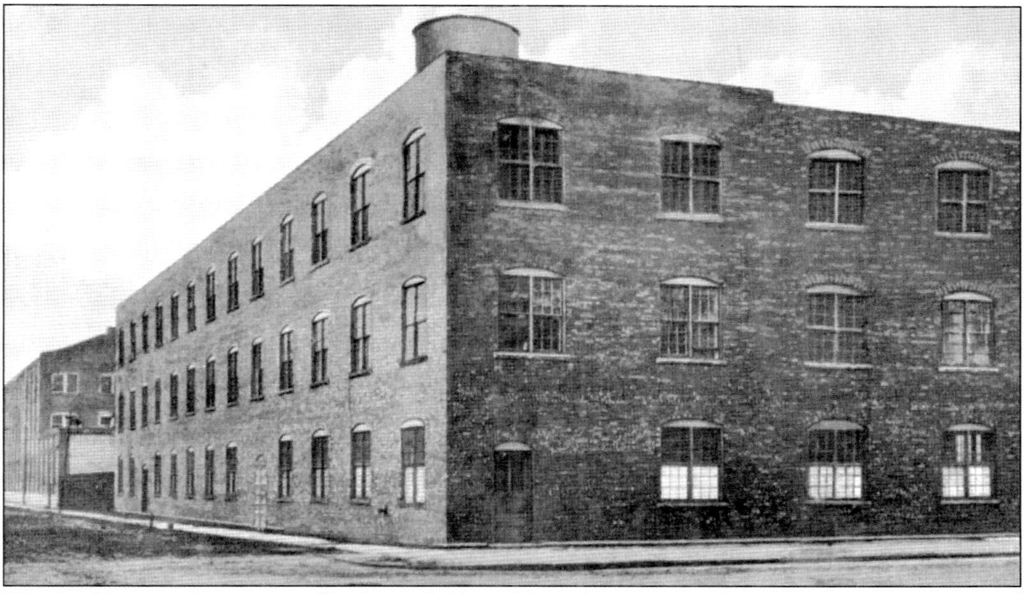

This building at the corner of Jackson (left) and Broad Streets was home to Conneaut Gas and Electric Fixture Company (1909), Conneaut Metal Works, Equipment Manufacturing Company (1919), Walter A. Goldsmith Company (1926), and The Astatic Corporation (1944).

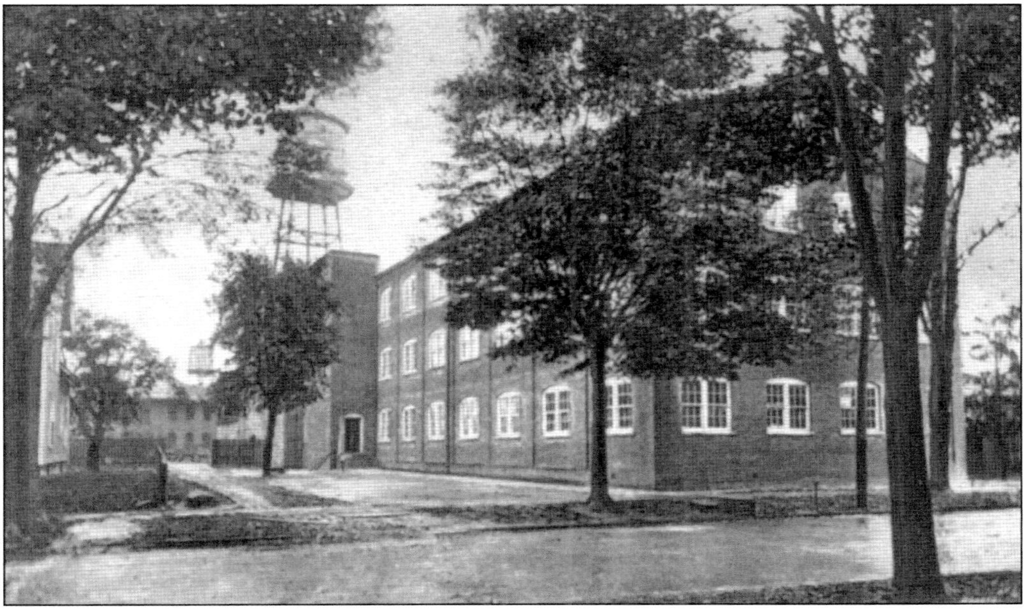

This building at the southwest corner of Harbor and Jackson Streets has been home to Conneaut Tungsten Lamp Works (1909), Conneaut Woolen Mills (1919), Stevens Yarn Company (1930), and The Astatic Corporation (1944).

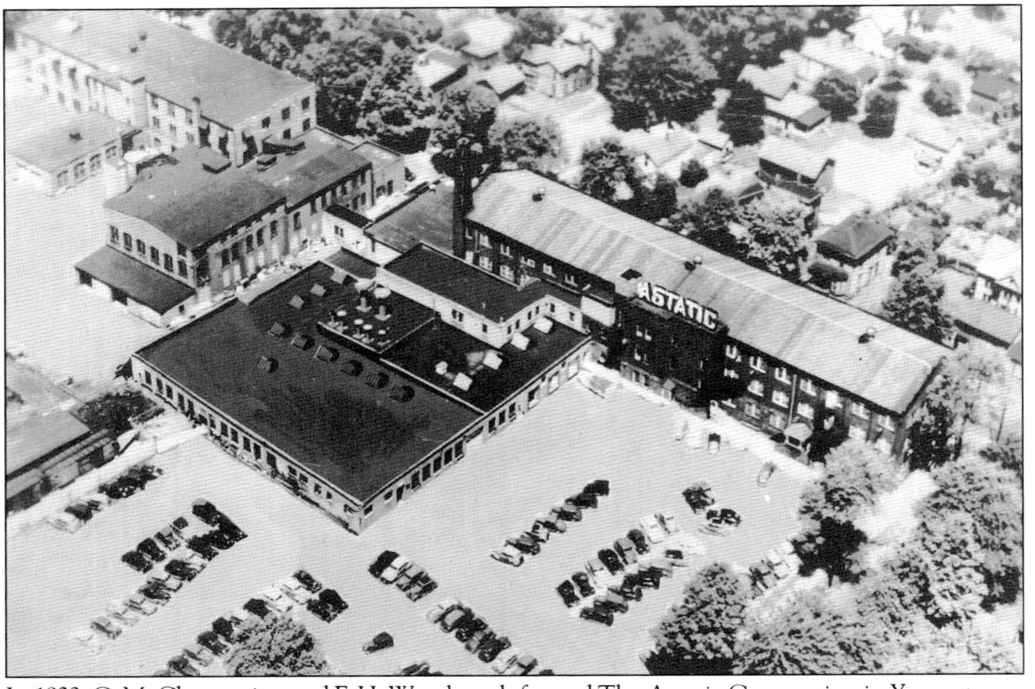

In 1933, C. M. Chorpening and F. H. Woodworth formed The Astatic Corporation in Youngstown, Ohio, to manufacture microphones for ham radio operators. In 1944, they moved the company to this 3-acre site bounded by Broad (left), Jackson (center) and Harbor (right) Streets. At one time, the company employed 1,200 people in the design, manufacture, and distribution of microphones and phonograph and tape-recording components. (Karl A. Rowbotham.)

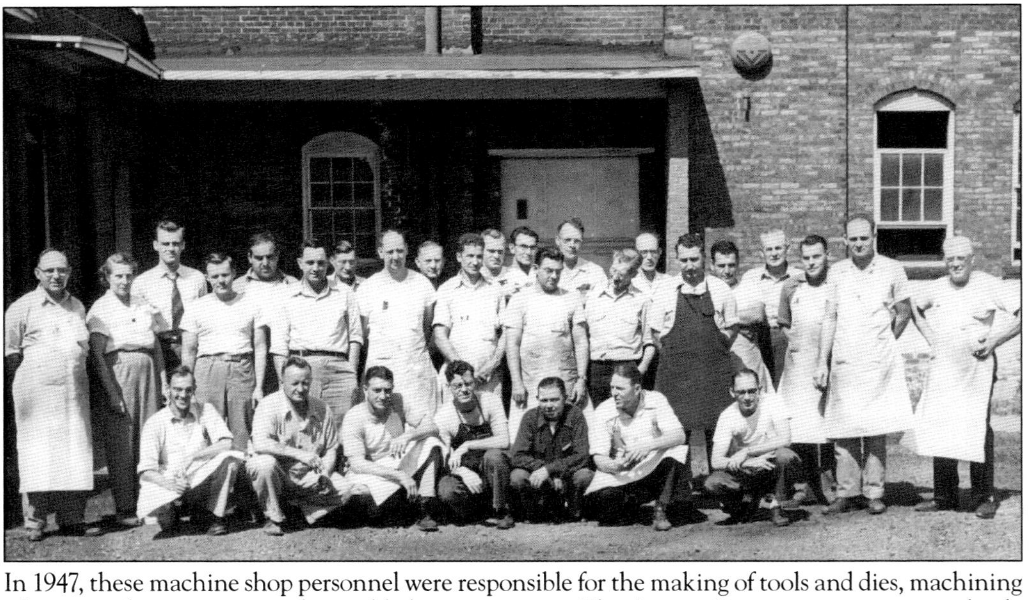

In 1947, these machine shop personnel were responsible for the making of tools and dies, machining of parts, and maintenance of assembly-line equipment. The Astatic Corporation was a completely self-sufficient business that was able to transform raw materials into finished goods with worldwide distribution from this one location. (Judith Owens Kreiling.)

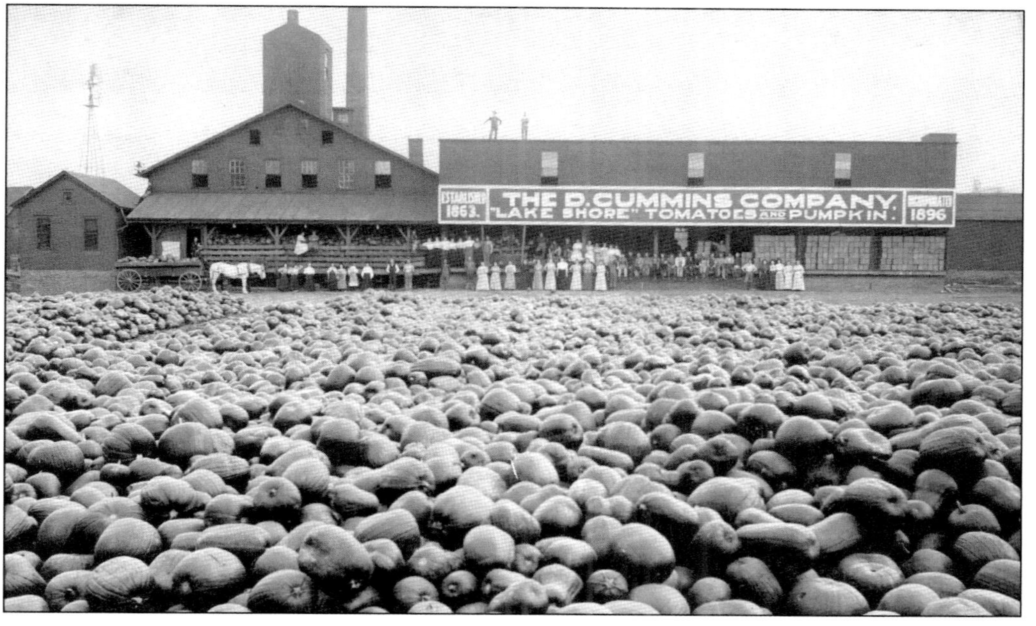

This c. 1900 photograph of the Cummins cannery was taken by a photographer facing south at today's intersection of Main Street and Cummins Avenue. The majority of these employees were seasonal, working only three months each year. On March 3, 1909, all of these buildings were destroyed by fire caused by an electrical wiring short at a loss of $100,000. The above cannery was replaced by the one below, which was located on the north side of Maple Avenue between Rieg Avenue and Center Street. The Whitney Avenue concrete and steel footbridge over the Nickel Plate Railroad yards can be seen on the right.

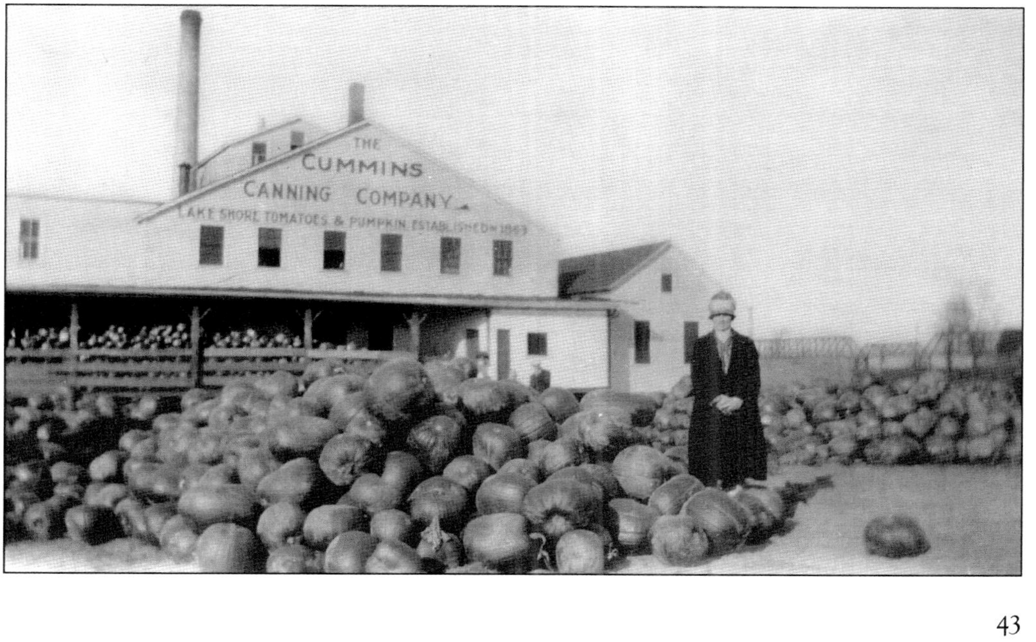

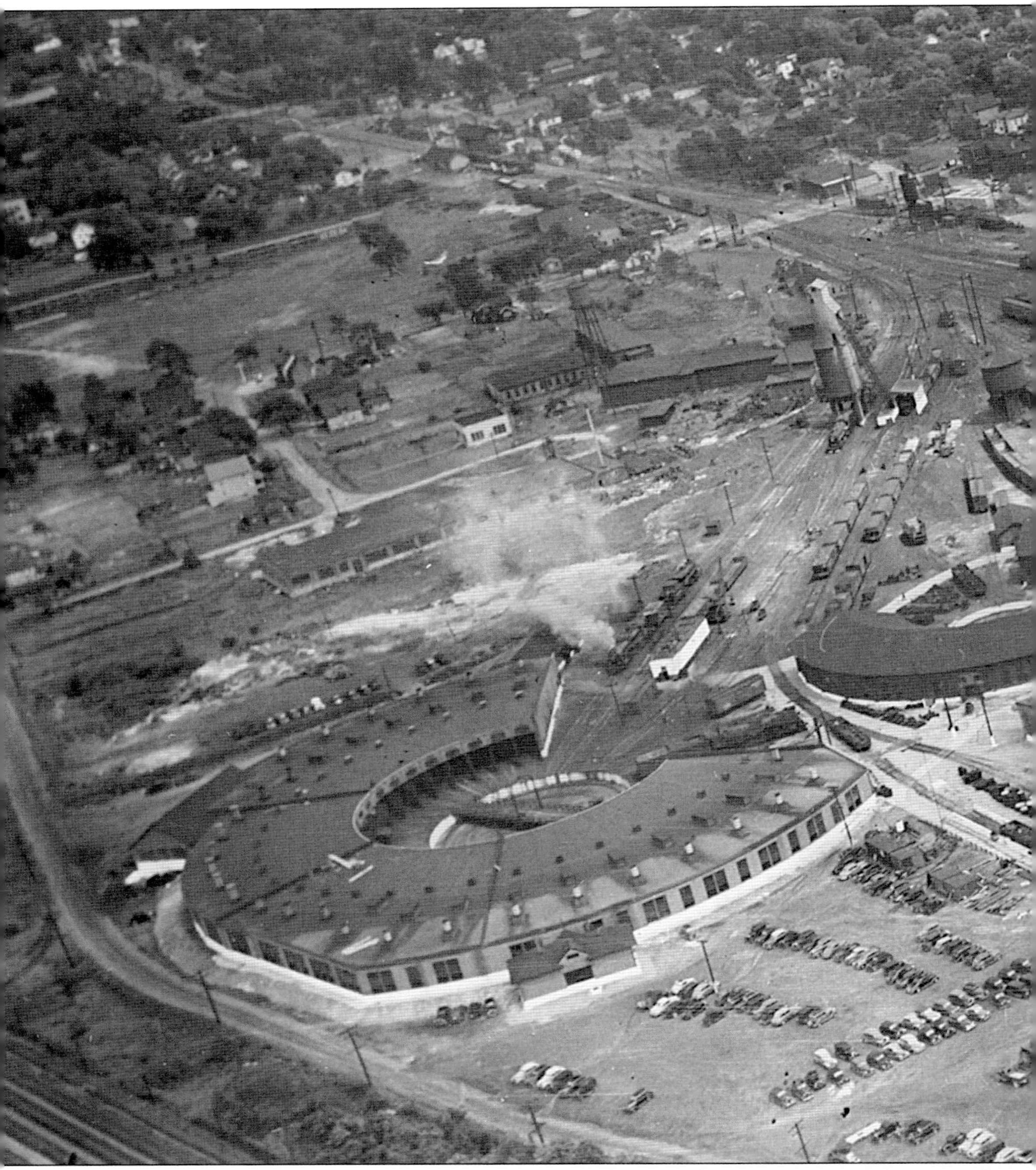

Conneaut High School and Conneaut Junior High School can be seen in upper left corner of this late 1940s photograph of the Nickel Plate Railroad yards. The large circular building to the lower left, with a turntable, is the roundhouse where steam locomotives were repaired. The tall, round structure just left of center with the locomotive under it is a locomotive water tower, which provided water for the tenders. The taller structure behind it is the sand house, which replenished sand for locomotive traction. Saint Mary's Catholic Church and Elementary School can be seen

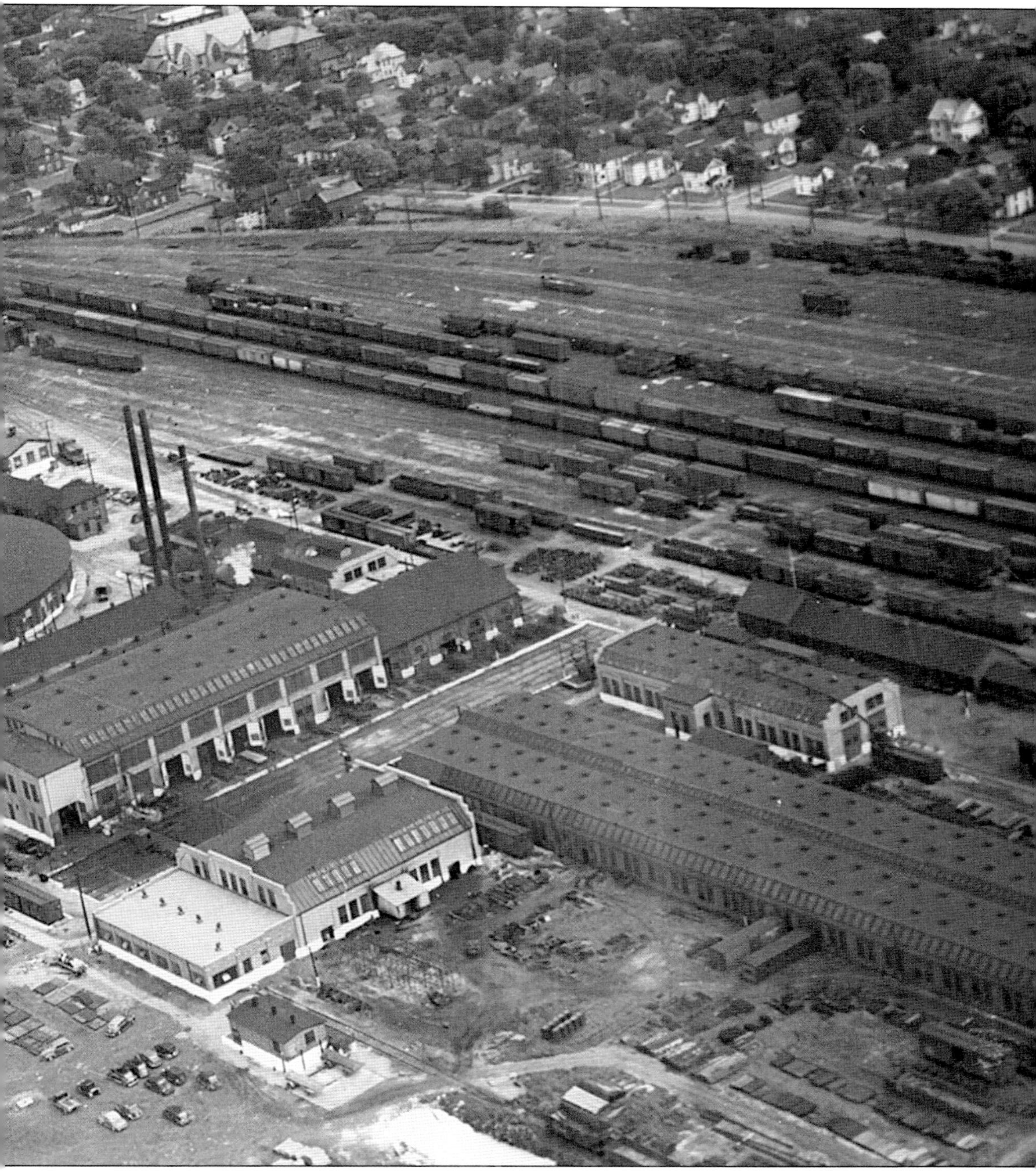

at the top, just right of center. The longest building to the lower right is the car repair shop, where repairs to rolling stock are performed. The buildings surrounding it include the boiler shop (lower left), machine shop (left), and woodworking shops (above right), which provided support for the roundhouse and the car repair shop. In the middle to upper right corner is the rail yard, where railroad cars are sorted and rerouted.

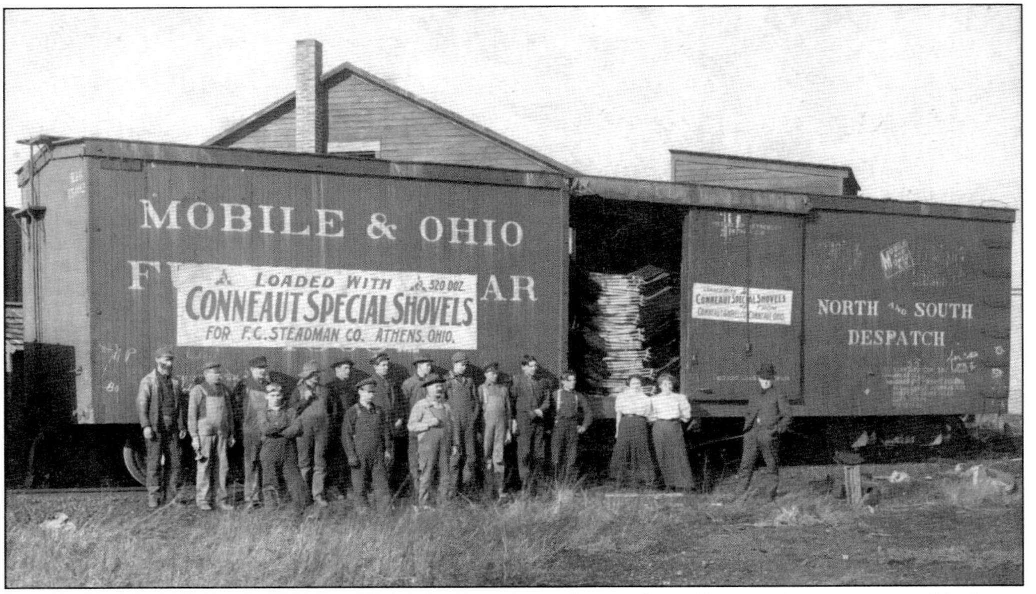

Conneaut Shovel Company, established in 1905, was located on the southeast corner of Jackson Street and Whitney Road. It was the largest shovel plant, under one roof, in the world. A radical innovation pioneered by the company was to design shovels for the environment in which they were to be used. In 1931, the company was acquired by American Fork and Hoe Company, which changed its name to True Temper in 1949.

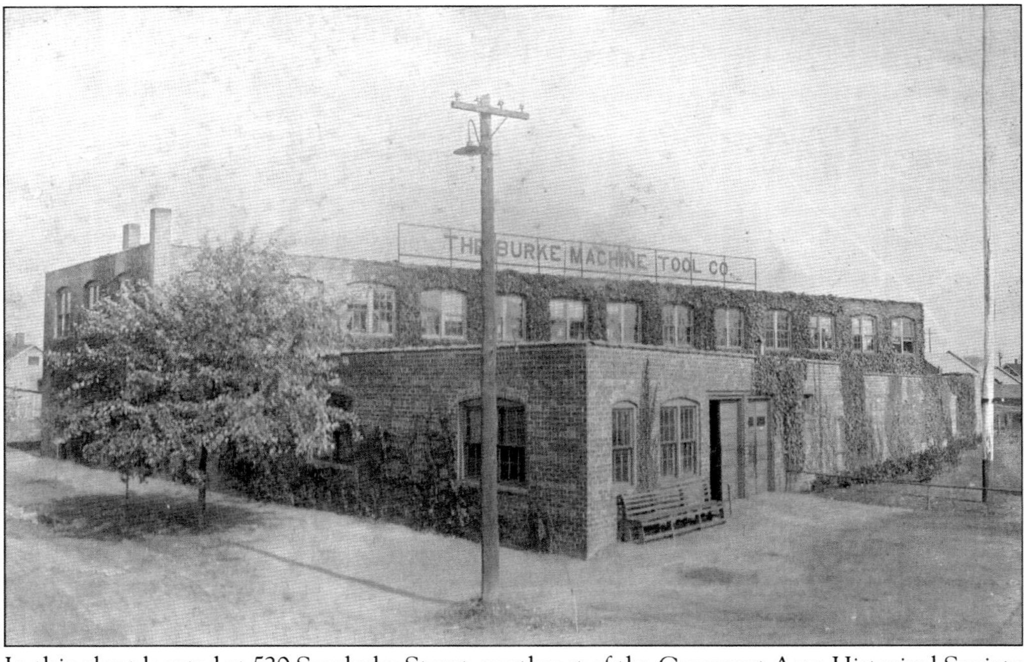

In this plant located at 520 Sandusky Street, northeast of the Conneaut Area Historical Society Museum, the Burke Machine Tool Company manufactured inexpensive milling machines and machinist drill presses from 1911 until 1948, when the company was purchased by Cincinnati Milling Machine.

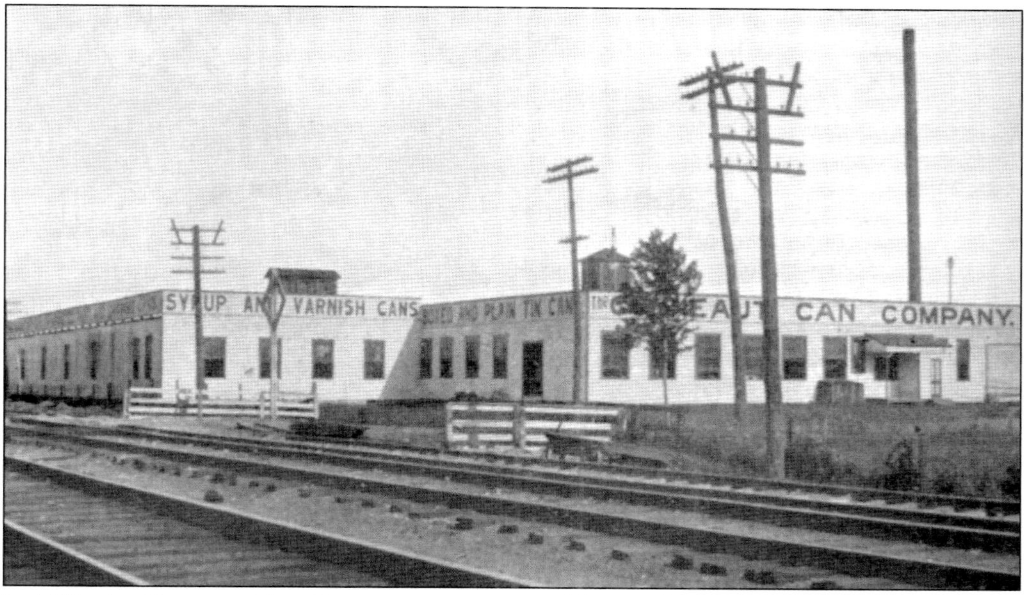

In 1901, the Conneaut Can Company was organized by two Record Manufacturing Company traveling salesmen when Record Manufacturing was acquired by American Can Company. H. A. Truesdale and Z. H. Thompson built their plant on the northwest corner of Jackson Street and Whitney Road. Between 1908 and 1935, the company provided lunch at no charge to employees in the company cafeteria. In 1956, Conneaut Can Company was acquired by Davies Can.

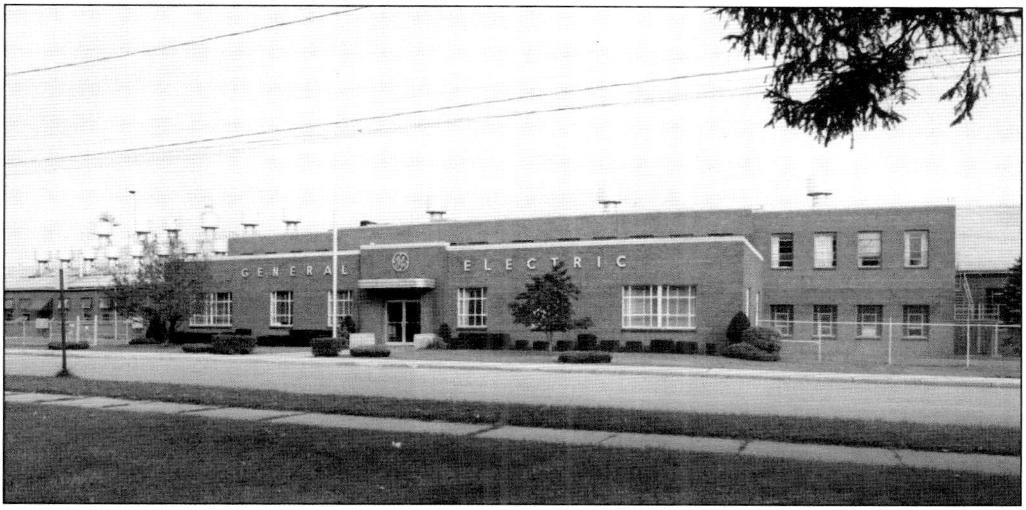

In July 1941, the General Electric Company's base plant opened at 800 Maple Avenue. During its 67-year history, it made light-bulb bases for incandescent, fluorescent, and automobile lights. Additional products included Christmas tree lights, MagiCubes, and Flipflash. At the factory's peak, it employed over 400 workers.

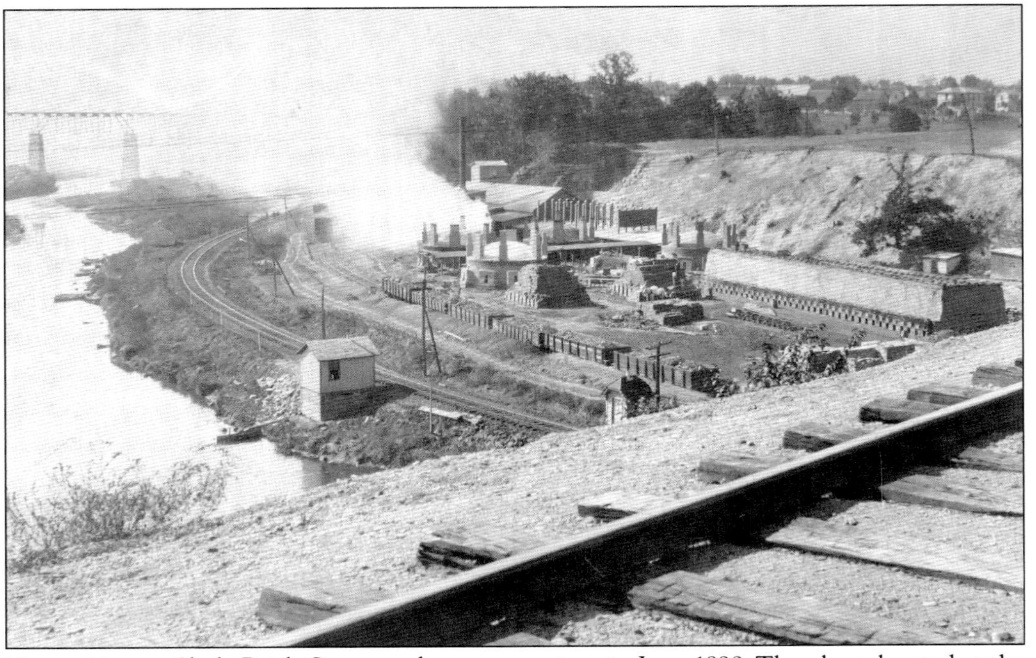

Harper-Norton Shale Brick Company began operations in June 1899. The plant, located at the foot of Marshall Street, had a daily capacity of 50,000 common and 22,000 repressed bricks. The round structures are beehive kilns, and the long structure to the right is a eudalay kiln. This photograph was taken from the top of the New York Central Railroad arches. The Nickel Plate Railroad trestle over Conneaut Creek can be seen to the upper left.

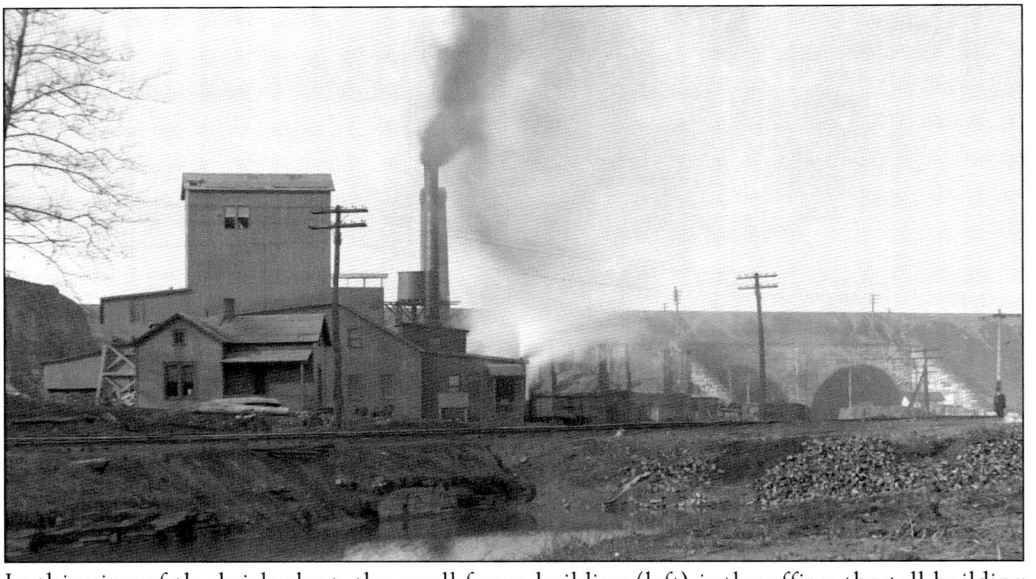

In this view of the brick plant, the small frame building (left) is the office; the tall building houses the shale grinding, clay mixing, brick extrusion, and cutting operations. The plant manager, Frank B. Dunn, devised the Dunn wire-cut lug process, which eliminated the repressing step for manufacturing paving bricks at this facility and its competitors. The plant ceased operations in 1924.

Four

PUBLIC SERVICES

In the late 19th century and early 20th century, immigrants flowed into Conneaut seeking jobs and a better way of life. The majority of the immigrants were from Finland, Italy, and Hungary. They worked primarily for the Nickel Plate, the docks, and building Conneaut's infrastructure. It was a common experience to arrive in Conneaut from the old country or Appalachia without a job and be working for the railroad or at the docks the next day. As the population mushroomed, the need for public services increased.

One of Conneaut's strengths has been its natural resources. Natural gas was first discovered in the village in 1852. Eastern Ohio Oil and Gas Company, headquartered in Conneaut, provided natural gas to customers throughout Ashtabula County. A competitor, Conneaut Gas, Light, and Fuel Company produced gas from 1889 until around 1905 at its plant on the north side of Conneaut Creek, midway between the Mill Street and Main Street bridges.

From 1895 to 1906, Conneaut's population nearly doubled, increasing from 5,186 to 9,647. With this increase in population came an increase in typhoid fever. The worst outbreak occurred in 1898 with 94 cases resulting in 12 fatalities.

In 1907, Conneaut's health officer, Dr. O. N. Warner, requested an investigation by the Ohio Department of Health. The investigation concluded that the typhoid fever outbreaks were caused by the public water supply, which was installed in 1891 by a private water company. After a few years, the venture proved to be unprofitable, and the company went into receivership.

The city water supply was drawn 1,500 feet from the shore at the foot of Broad Street, as shown on the 1912 Conneaut Harbor Chart on page 15. The city's raw sewage was not treated before being dumped into Conneaut Creek. The prevailing westerly winds normally forced the water from Conneaut Creek to flow to the east. However when the winds came from the east, pathogens from the untreated sewage in Conneaut Creek would be drawn into the city's water supply. Furthermore, because of the financial condition of the privately owned water works, the chemicals used to treat the water were often in short supply, and proper procedures were not followed.

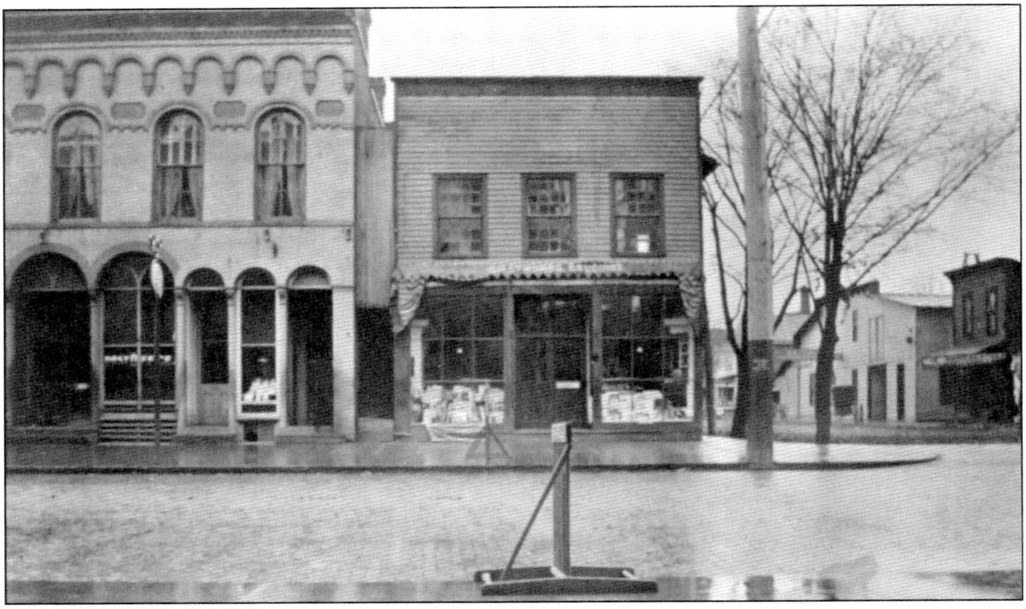

In 1808, Western Reserve postal service began when John Metcalf was hired to carry mail on foot between Erie and Cleveland. By 1812, he was using a horse-drawn lumber wagon. In 1815, mail was delivered to Conneaut by stagecoach twice each day. In this 1890 photograph, the post office was located at 225 Main Street (left).

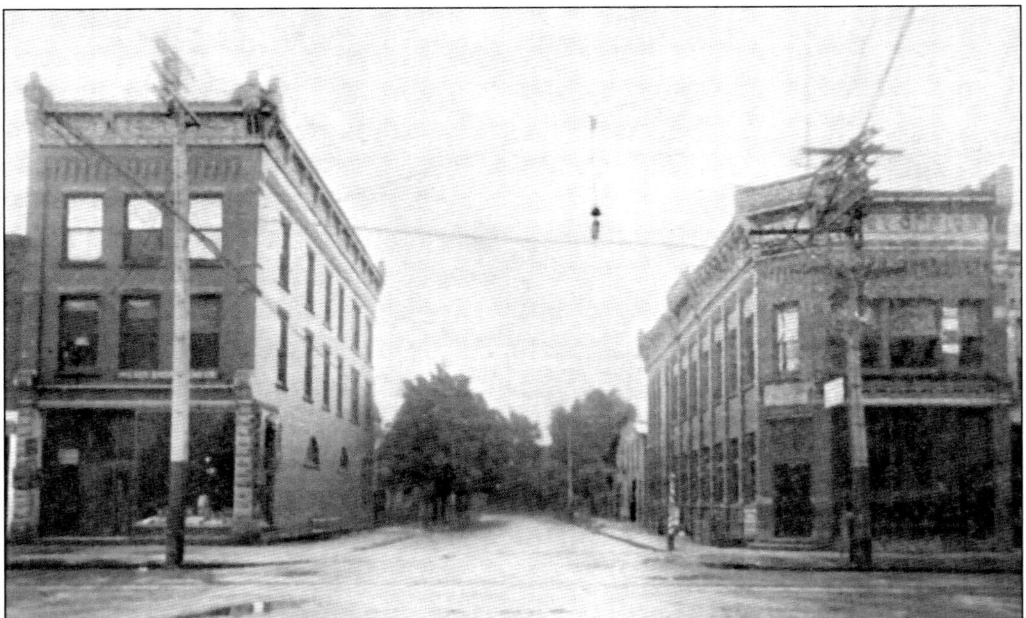

By 1902, the post office had moved to 233 Main Street on the southwest corner of Washington Street. The frame building on the opposite side of the Washington Street has been replaced by a three-story brick structure. J. M. Burhan's livery stable behind the post office houses the fire department's horse-drawn pumper.

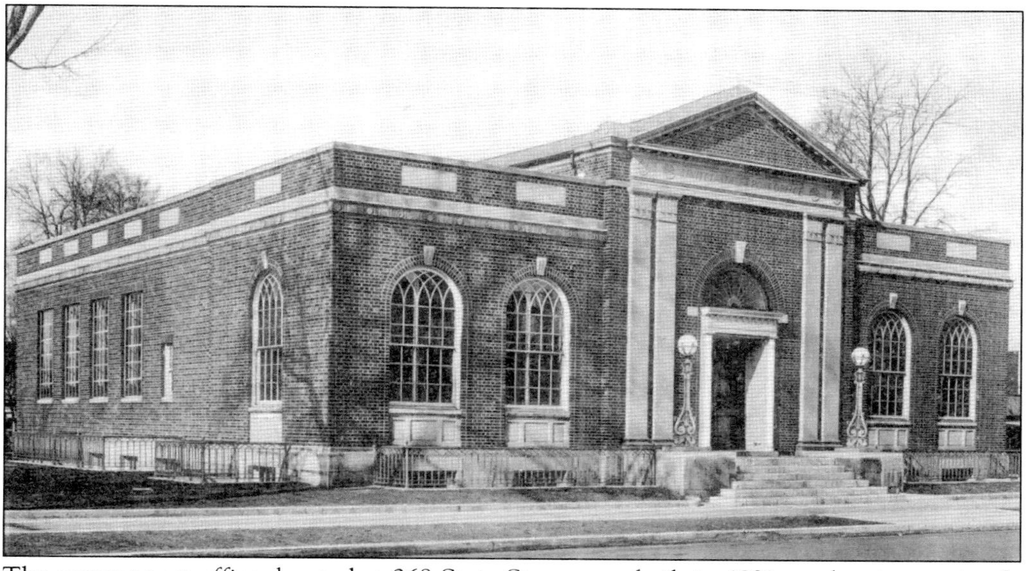

The current post office, located at 268 State Street, was built in 1931 on the site previously occupied by a filling station. The window service lobby was in the center and post office boxes in the right front corner of the building. Military recruiting stations were located in the basement. During the 1920s, the post office was located at 223 Washington Street.

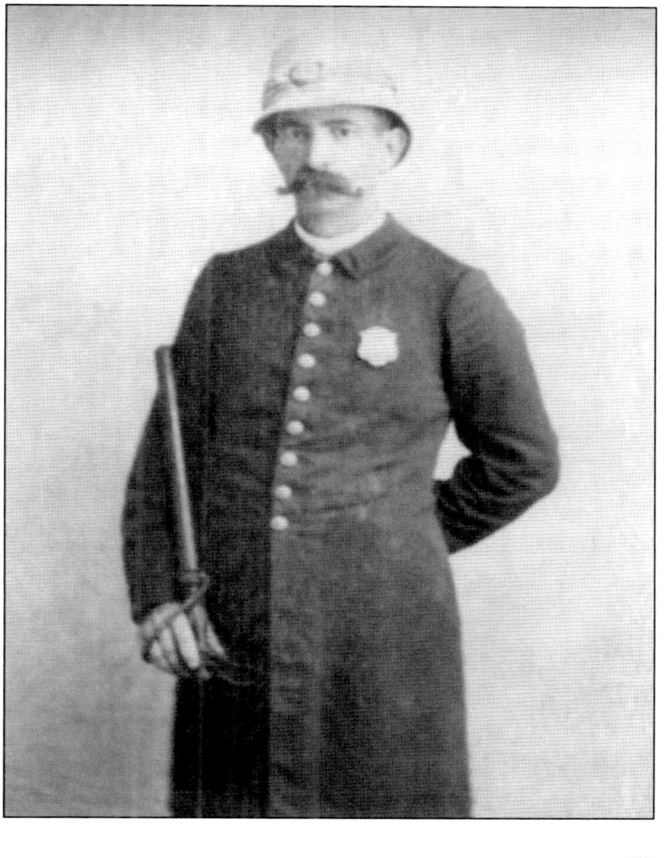

George J. McMillen became Conneaut's first policeman in 1892 and was elected marshal in 1893. McMillen served as chief of police under three mayors: Sagito J. Smith (1897 to 1899), C. Henry Simonds (1899 to 1901), and Wilber F. Stanley (1901 to 1903). In 1907, he retired from the Conneaut police force with approximately 15 years of service. (Conneaut Police Department.)

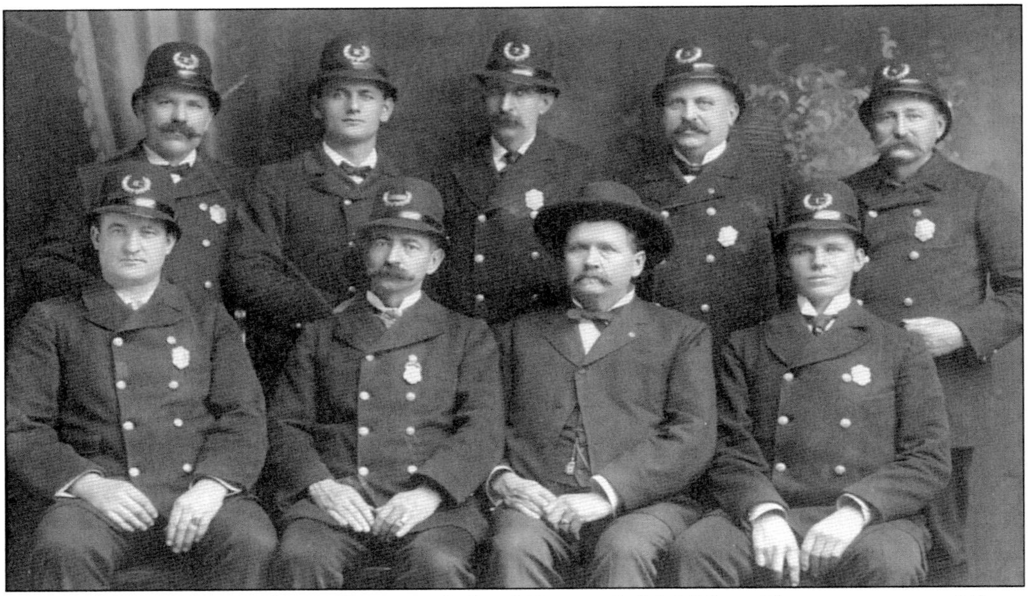

In 1900, Conneaut's police officers and chief George J. McMillen (front row, second from left) pose with Mayor Wilber F. Stanley (front row, third from left). At this time, horses were restricted to a 5 mile per hour speed limit within the city. In 1920, the first police car was purchased. Prior to becoming radio equipped in 1951, red lights mounted on utility poles were used to communicate with officers. An illuminated light meant the officer must contact the station. (Conneaut Police Department.)

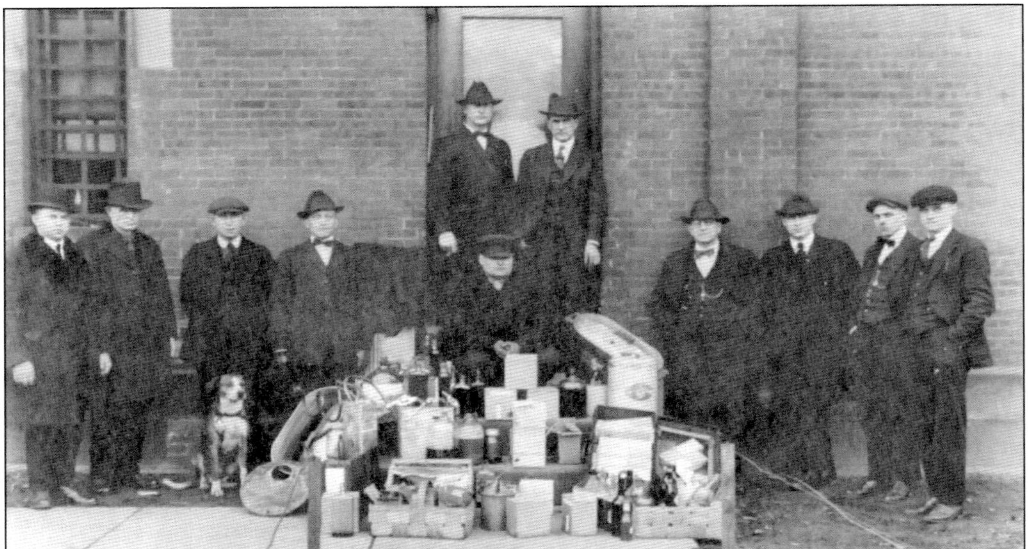

In 1922, the police department displays contraband confiscated during bootleg raids. Organized crime, known as the Black Hand, flourished in Conneaut during Prohibition. Under the cover of darkness, rumrunners regularly smuggled liquor across Lake Erie from Canada. The Harbor Bombers mob gained notoriety attacking homes and businesses with sticks of dynamite. In August 1930, newlywed Anthony Roberto was murdered because the mob feared he might testify regarding their failed bombing of a former city councilman's home. (Conneaut Police Department.)

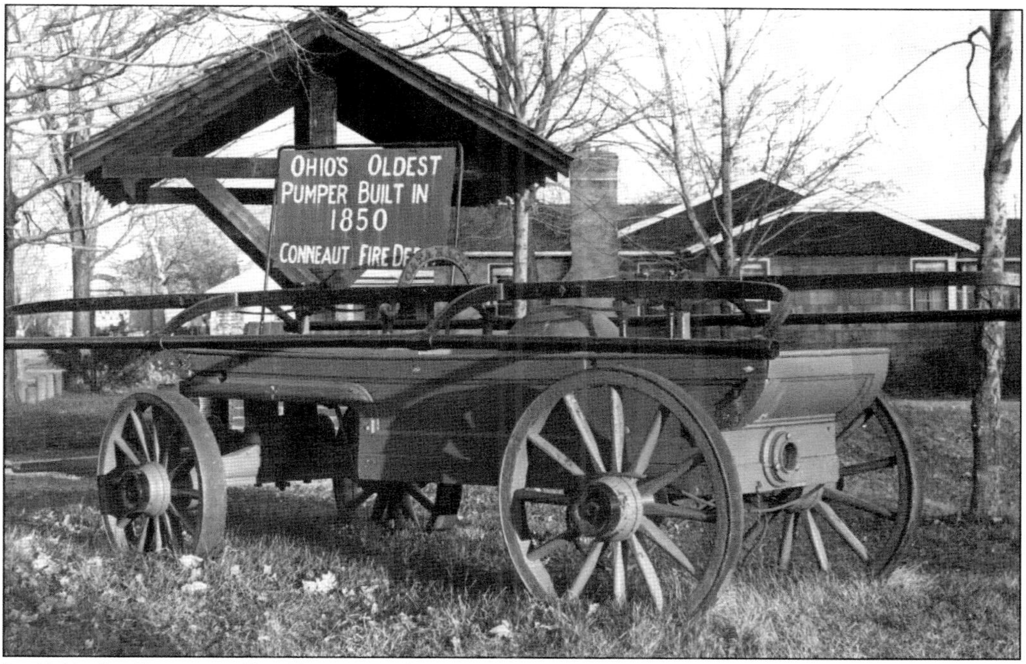

In 1841, the Conneaut Fire Engine and Hose Company, a 21-man volunteer group, was organized. In 1850, the group reorganized as the Cataract Fire Company with 40 volunteers. At that time, this 22-man pumper was purchased for $1,000. In the event of a fire, water was drawn from public wells or cisterns.

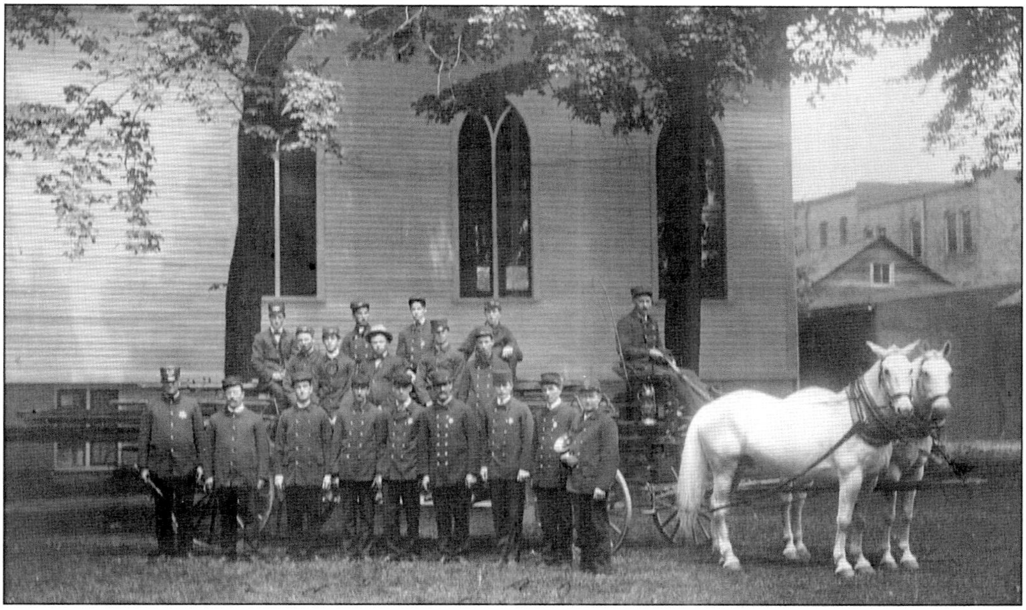

The Hook and Ladder Company was organized in 1892. The company is shown next to the Christian church behind the town hall in this photograph. Although the company had a reputation for being responsive, their equipment was lacking. This was evidenced by the Conneaut Paper Mill fire, which was considered under control until the fire hose burst. (Conneaut Fire Department.)

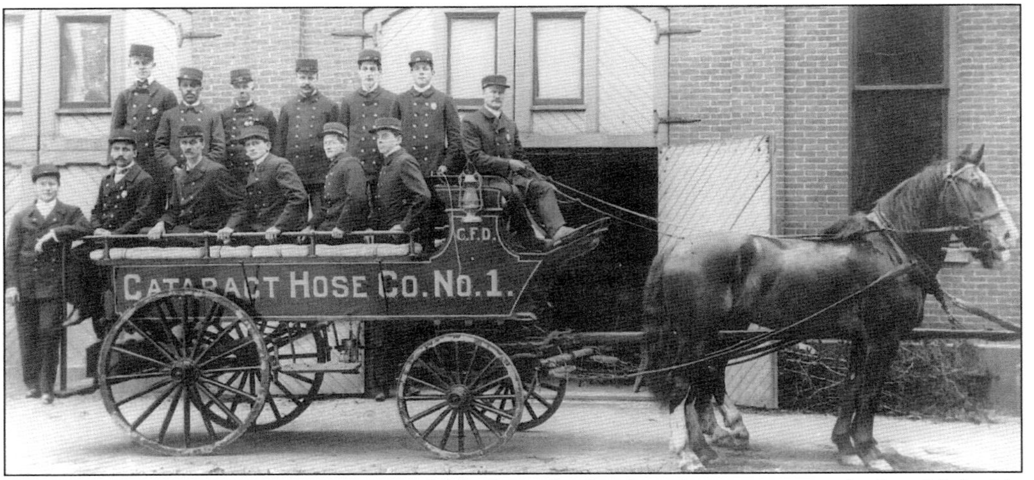

By 1893, the city had Lake Erie water available through 13 miles of cast water pipe feeding 36 double fire hydrants. However, fire wells and cisterns were still used in parts of the city. Around this time, the Cataract Company acquired this horse-drawn hose wagon, and the Niagara Hose Company was organized to furnish fire protection to the west end. (Conneaut Fire Department.)

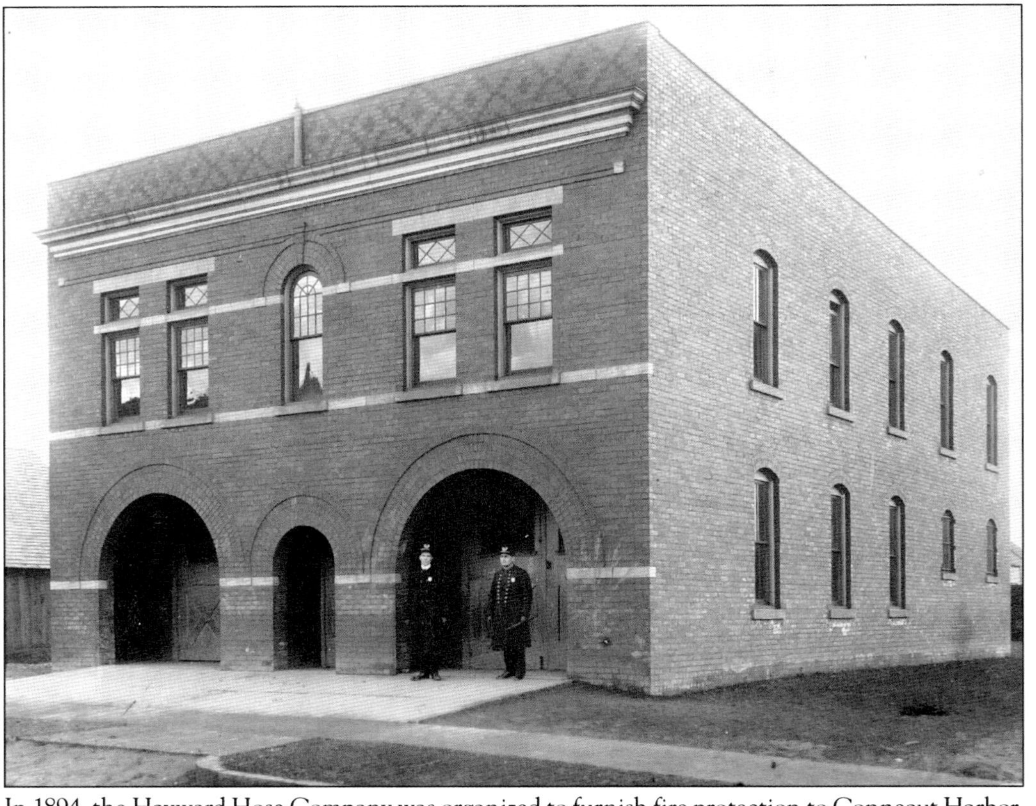

In 1894, the Hayward Hose Company was organized to furnish fire protection to Conneaut Harbor. By 1913, the city had a salaried fire department headquartered at city hall and a harbor fire and police substation (pictured) at 185 Erie Street. When the city purchased its first motorized fire truck, this station became the home of the 1893 Cataract hose wagon.

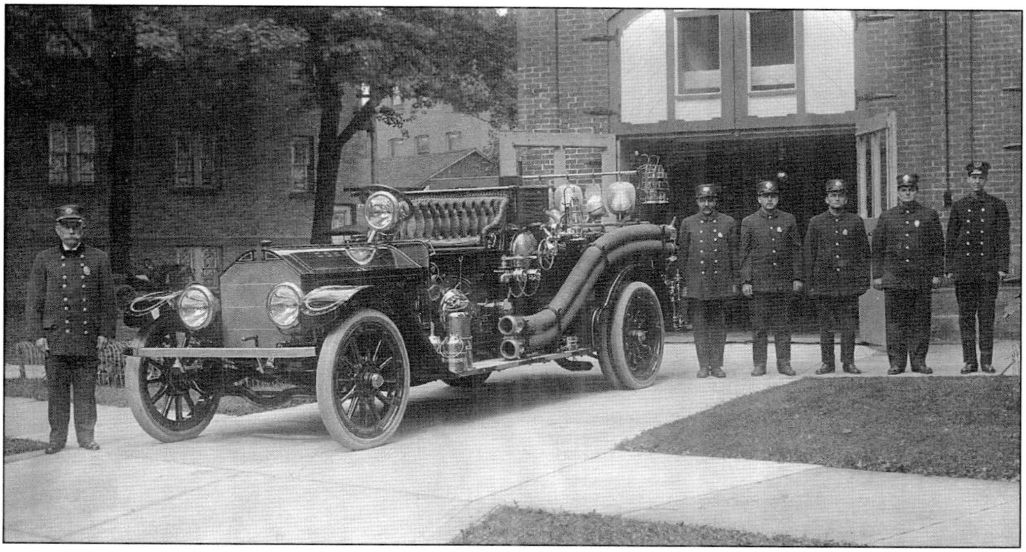

In December 1913, an American LaFrance combination hose chemical and water pumper (above) was purchased for $9,000. Chief Thomas J. Gough (left) poses with the first salaried firemen. The same year, a Gamewell fire alarm system was installed throughout the city. The system consisted of 42 boxes (below, left), which were mounted on utility poles and marked with a wide red band between two narrow white bands. When an alarm box lever was pulled, a bell at the station (below, right) sounded, which indicated the alarm box number. The system was discontinued approximately 50 years later, when telephones in every home became the norm. (Both Conneaut Fire Department.)

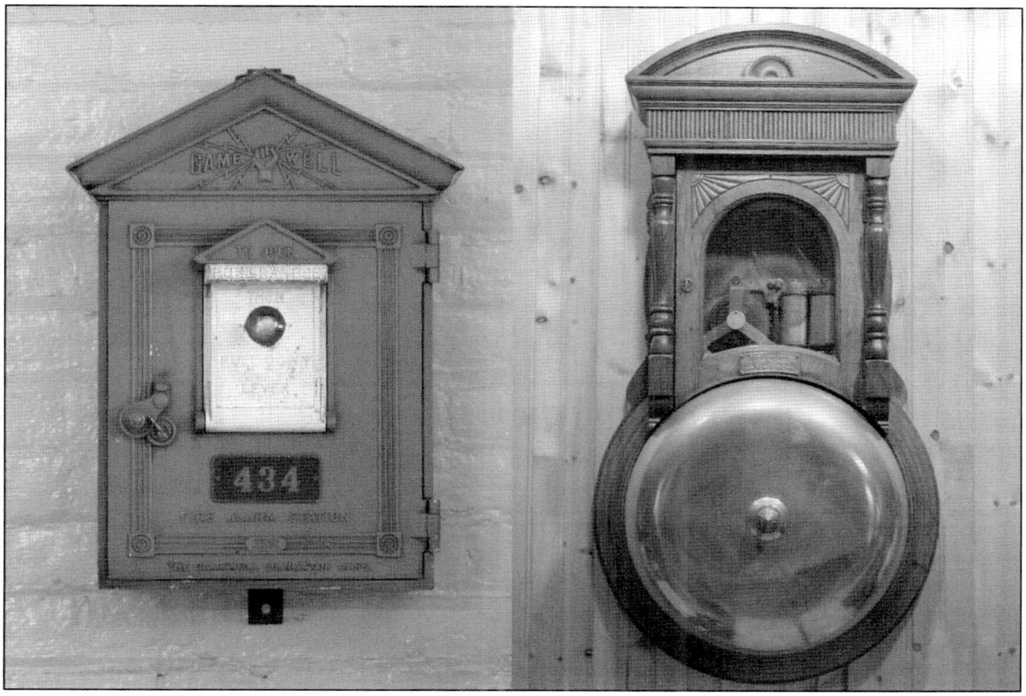

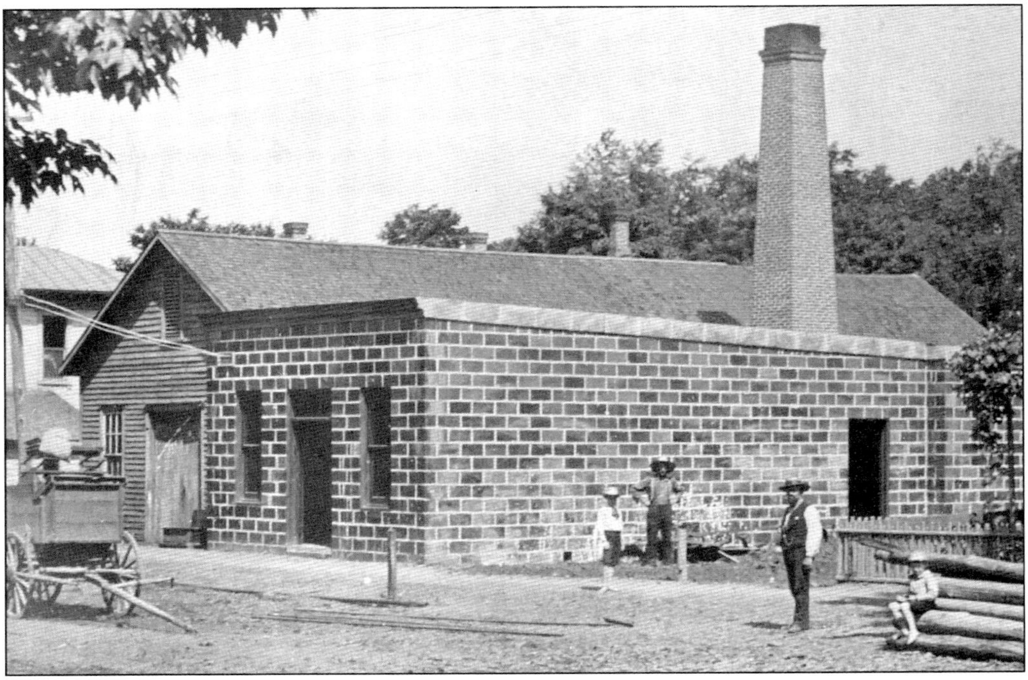

Conneaut's first electricity was generated by the Webb Incandescent Electric Light Plant from this building at 318 State Street. William H. Webb appears in this photograph with his sons, Charles, Irwin, and Carley. The capacity of the plant was 310 incandescents, approximately 15,500 watts, nearly enough to provide electricity to 12 average American homes in 2001.

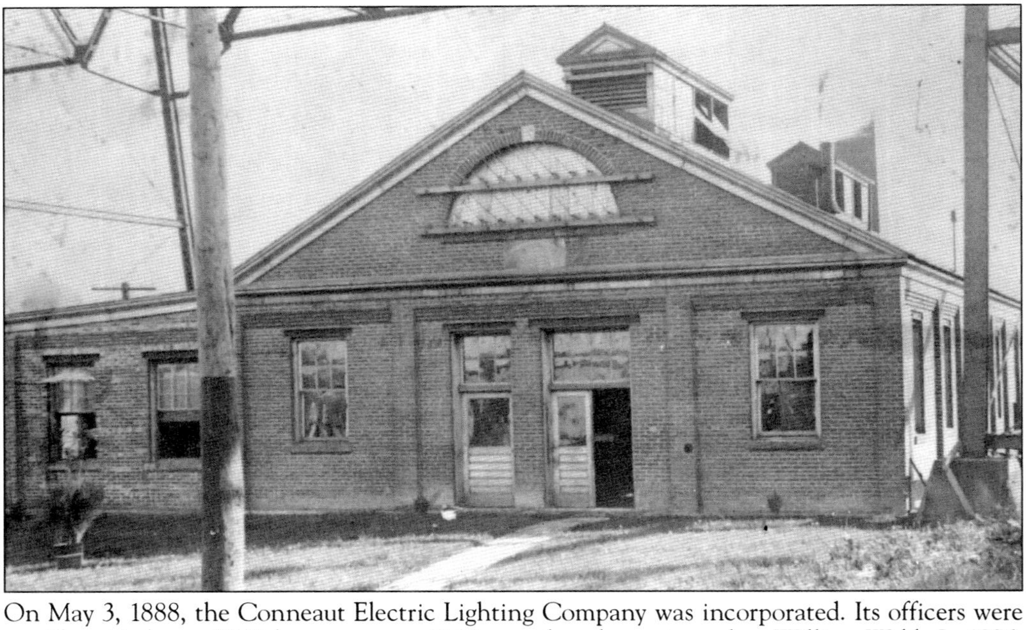

On May 3, 1888, the Conneaut Electric Lighting Company was incorporated. Its officers were president Cushman W. Pelton, secretary C. Hayward, and superintendent William Webb. In 1896, this new Conneaut Electric Light plant was completed at 125 Main Street. In 1926, the plant was purchased by the Cleveland Electric Illuminating Company and then closed.

This pole at 214 Washington Street brought all the telephone lines into the Conneaut Telephone Company's central office on the second floor of 228 Main Street. Edward L. Barber and James S. Brailey, of Wauseon, Ohio, installed telephone equipment throughout the eastern United States, acquiring stock in the local telephone companies. Edward L. Barber was the first president of the Conneaut Telephone Company when service was initiated in August 1897.

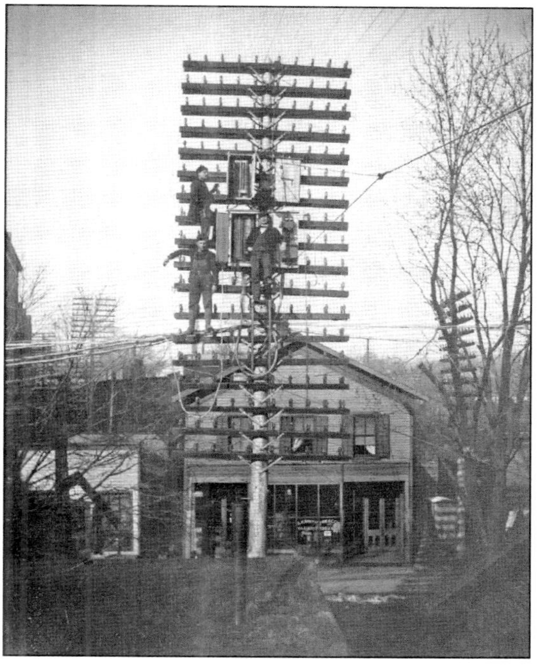

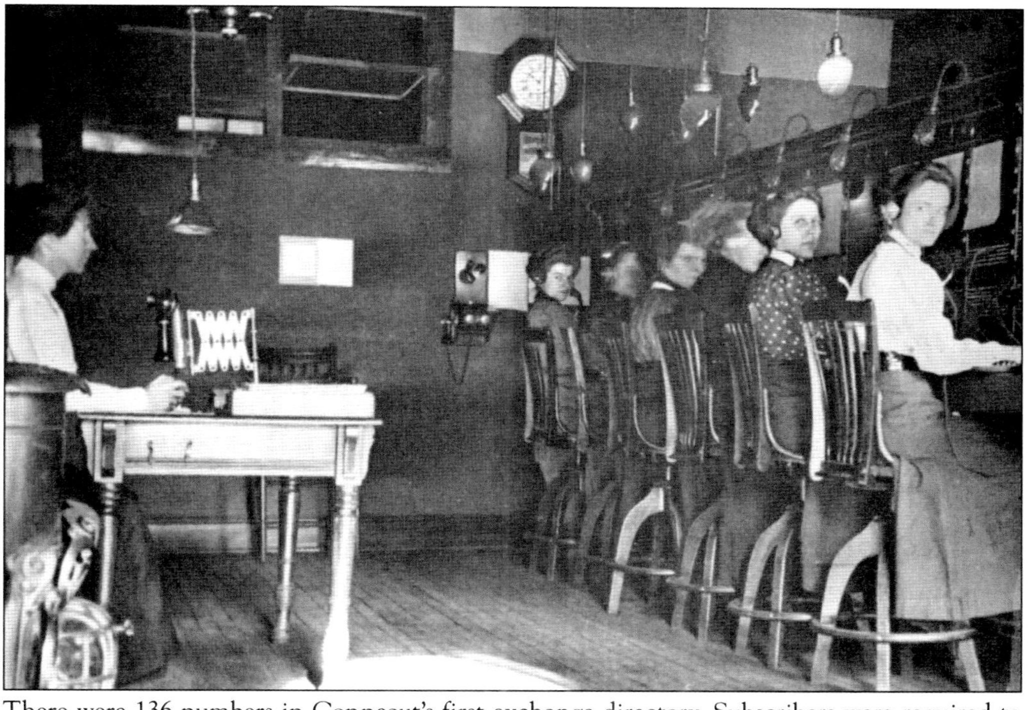

There were 136 numbers in Conneaut's first exchange directory. Subscribers were required to speak with an operator, who would connect them with the other subscribers. Service was available 24 hours a day, except Sunday, when it was available three hours in the morning and two hours in the afternoon. In 1912, the Conneaut Telephone Company eliminated the competition by acquiring the Central Union (later, Bell) Telephone Company's holdings.

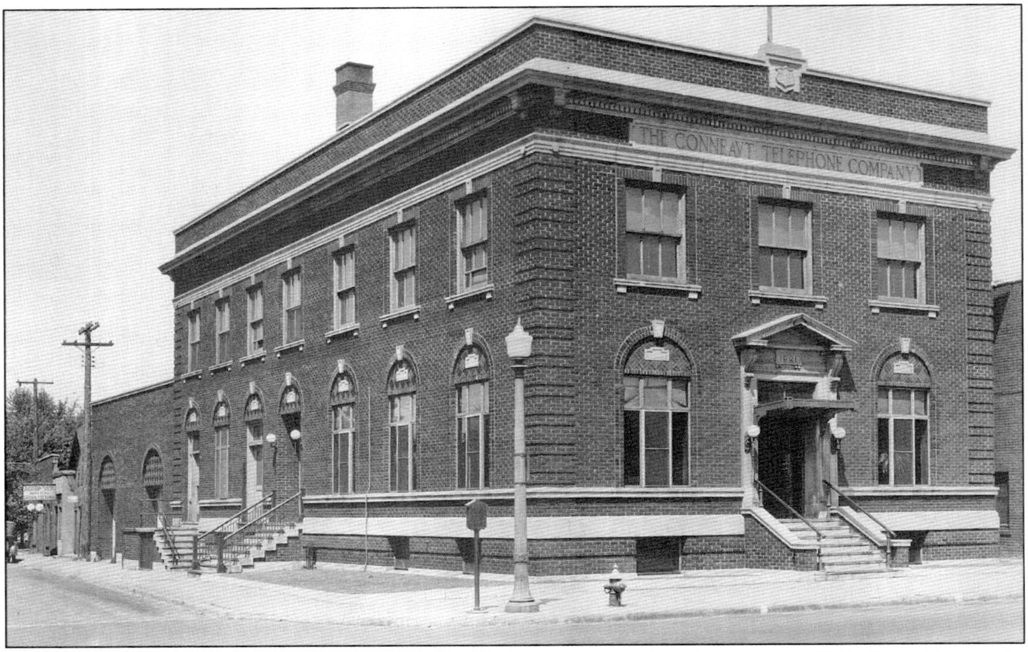

In 1921, the Conneaut Telephone Company built its current headquarters at 228 State Street. In 1941, the operator system was replaced with the rotary dial system, eliminating the operator and making the company one of the earliest to provide dial service to its subscribers. In 1966, it further expanded its facilities when it acquired and demolished the State Theater building (far right).

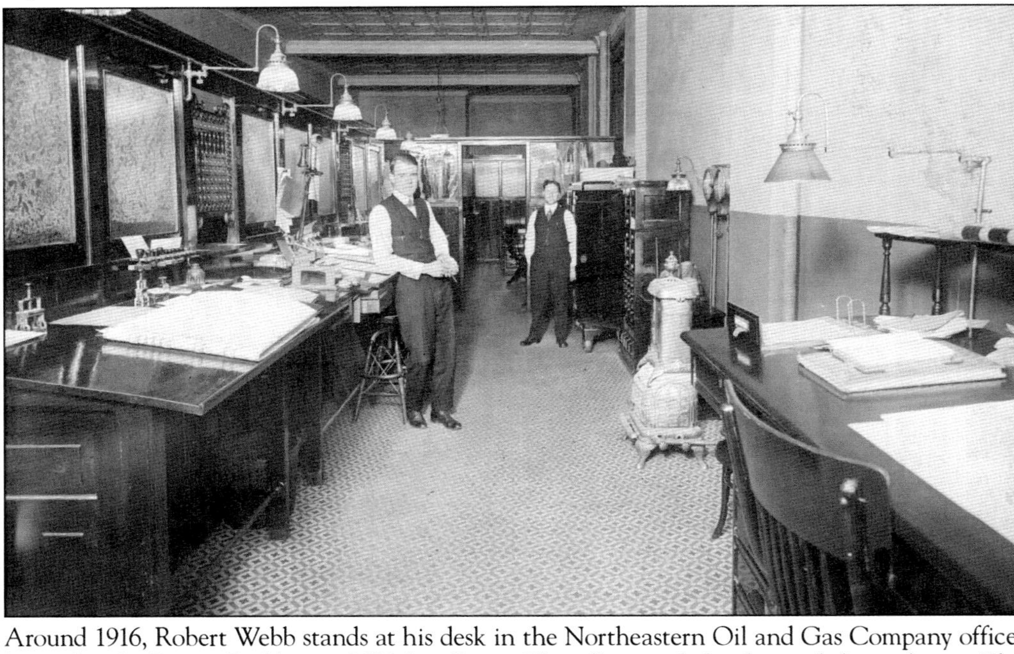

Around 1916, Robert Webb stands at his desk in the Northeastern Oil and Gas Company office, located in the Brown Building at 267 Main Street. The office was lighted entirely by gas lamps. The company, headquartered in Conneaut, had over $500,000 in assets, including 15 gas wells throughout Ashtabula County. The company president was Frank B. Dunn. (Joan Webb Bennett.)

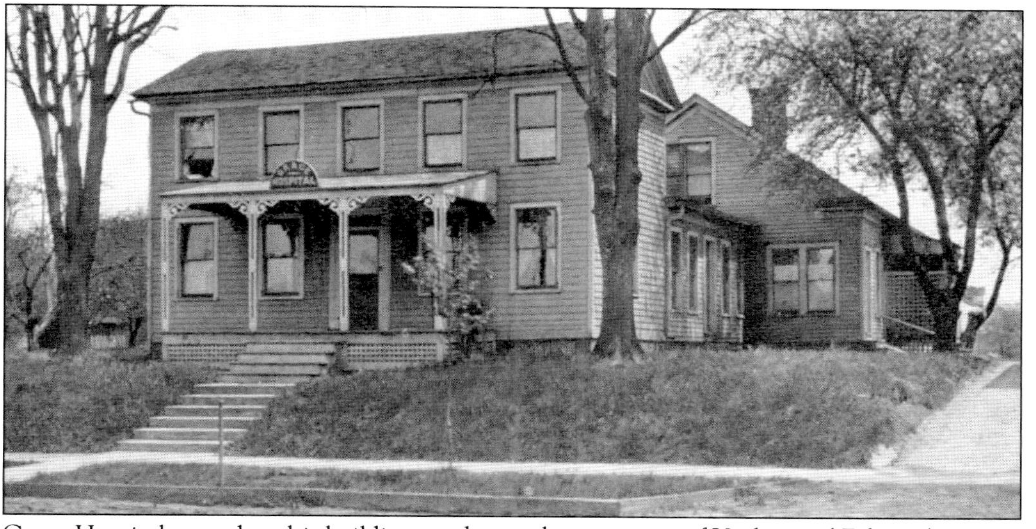

Grace Hospital moved to this building on the southwest corner of Harbor and Fifteenth Streets in 1905. The hospital, founded by two nurses in 1903, was also a nurse training hospital. In 1896, Dr. George H. Cole had opened Conneaut's first hospital at 311 Main Street. He described his 12-bed homeopathic facility as "a general hospital for the cure and treatment of all classes of cases with the exception of the insane." (Mary Jane Fritts.)

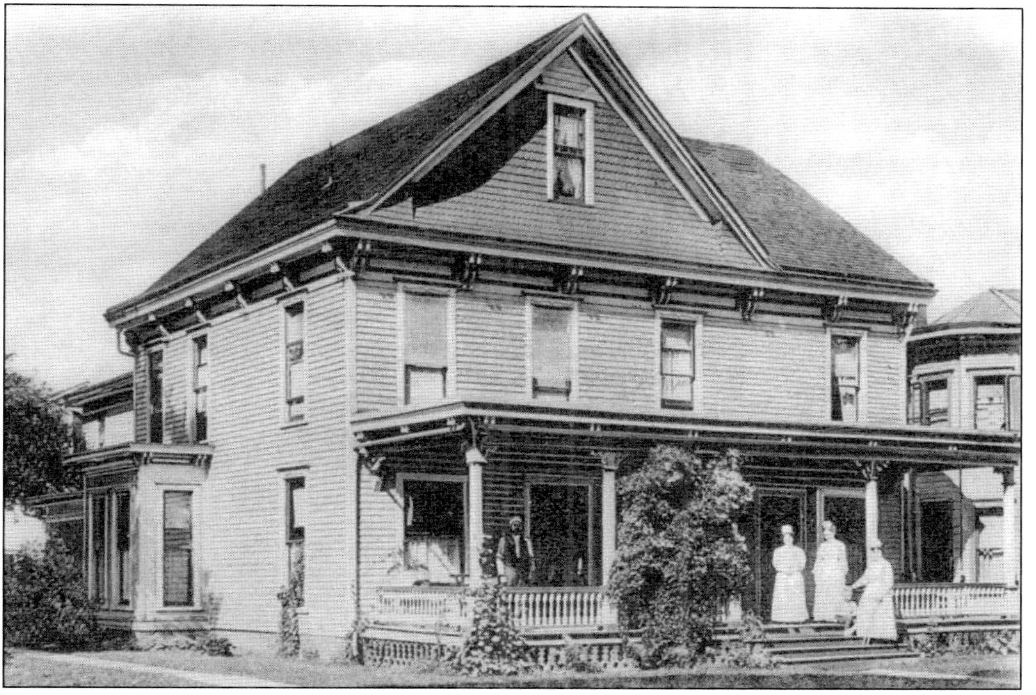

Conneaut General Hospital, located at 283 Harbor Street, was opened by Dr. Charles Everette Watson and his wife, Dr. Mabelle S. Watson, in August 1902. In January 1903, Charles died of tuberculosis; the hospital closed and his wife pursued post-graduate studies and went on to practice medicine in Ashtabula. May Record Findley, who lived next door, at 275 Harbor Street (left), would die of tuberculosis two years later at Grace Hospital. (Mary Jane Fritts.)

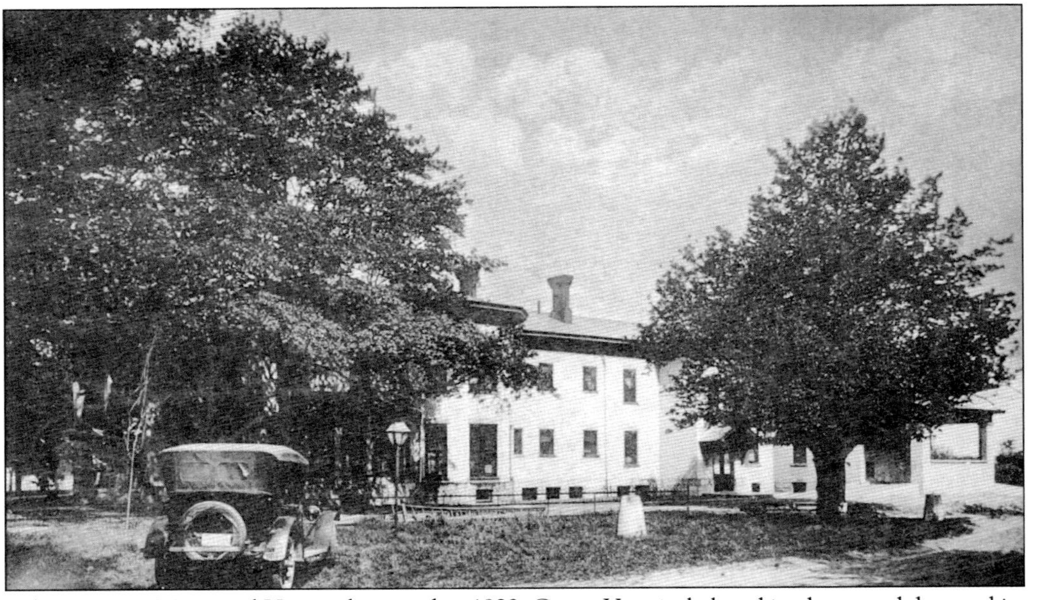

When Brown Memorial Hospital opened in 1922, Grace Hospital closed its doors and donated its equipment to the new hospital. George M. Brown, a prominent Conneaut banker and businessman, donated his property for use as a hospital upon his death in 1919. The community raised $80,000 for renovations, such as the addition of an emergency room and surgery rooms to the back of the building.

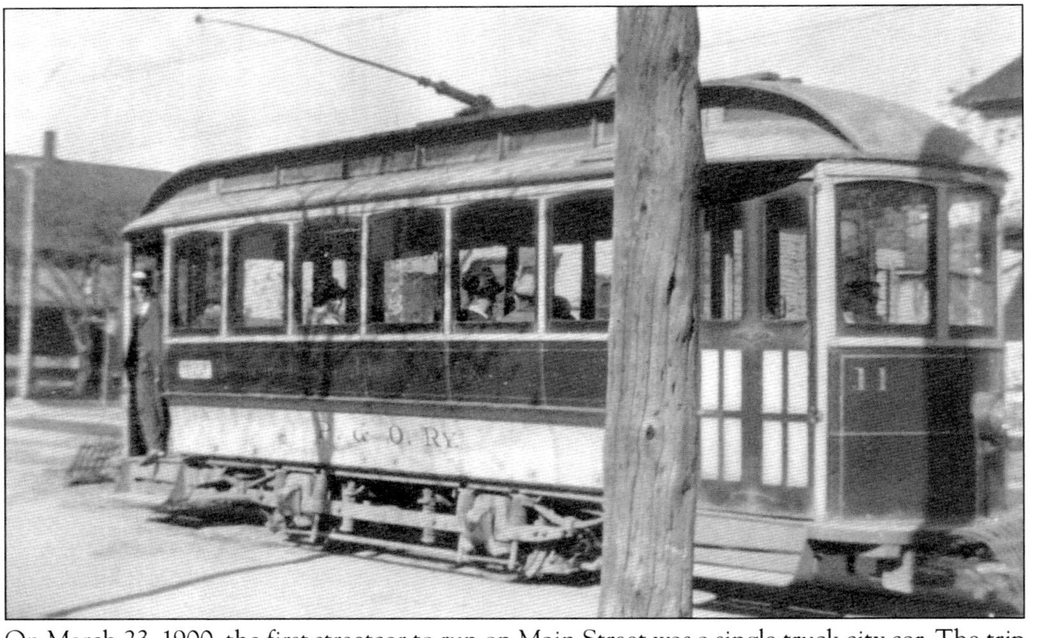

On March 23, 1900, the first streetcar to run on Main Street was a single truck city car. The trip began at the car barns, on the south side of West Main Street, near Townsend Avenue. The car headed east on Main Street to the terminal, north on Harbor Street, west on State Street, north on Sandusky Street, east on Depot Street, north on Broad Street, and east on Park Avenue to Day Street. (Kingsville Public Library.)

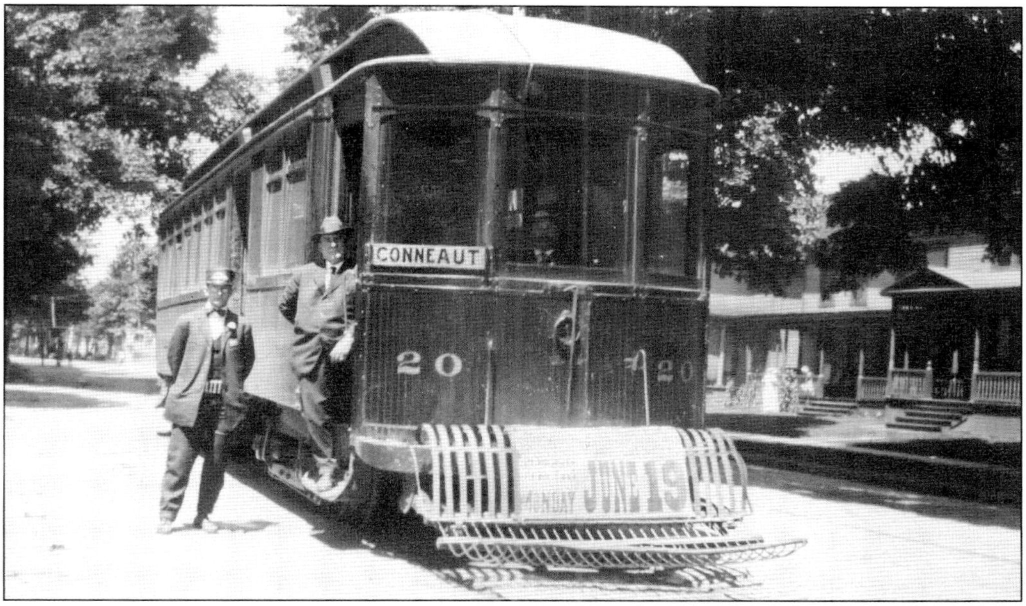

On November 11, 1901, the first Pennsylvania and Ohio Railway interurban trolley arrived at the West Main Street car barn from Ashtabula. On September 25, 1903, the first trip was made from the terminal at Harbor and Main Streets along the Conneaut and Erie Railway line to Erie. The trolley lines proved to be unprofitable and stopped running by February 24, 1924. (Kingsville Public Library.)

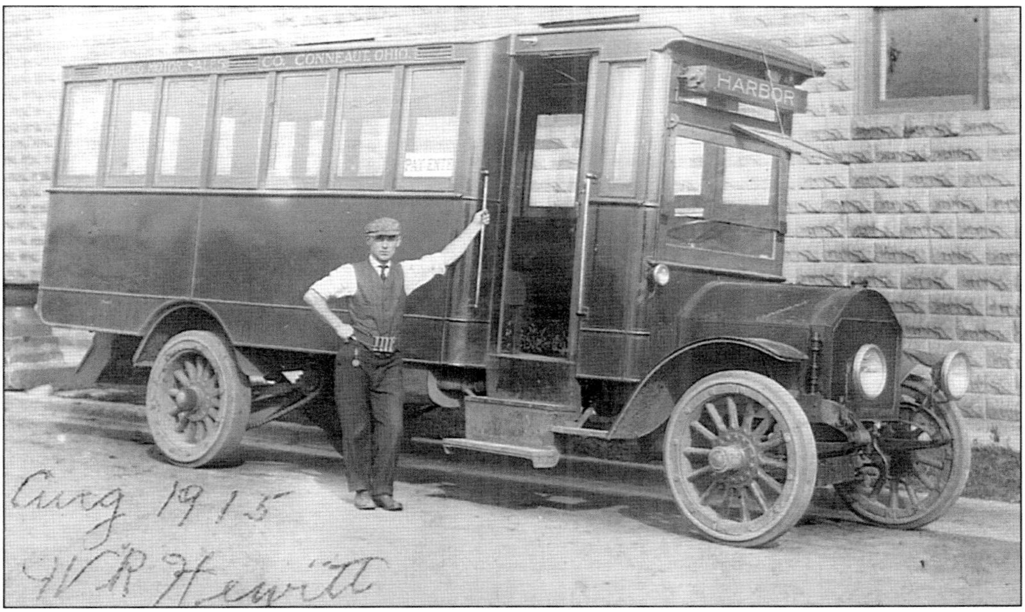

Early bus service, provided by independent jitney drivers, began in 1915 with the fare being a jitney (nickel). Jitney drivers followed no particular schedule and would make a U-turn to pick up a passenger headed in the opposite direction. In 1922, the city line trolleys stopped operating, and the Conneaut Bus Company was granted a license to operate buses within the city. City bus fare was 10¢ until they stopped operating in the 1960s.

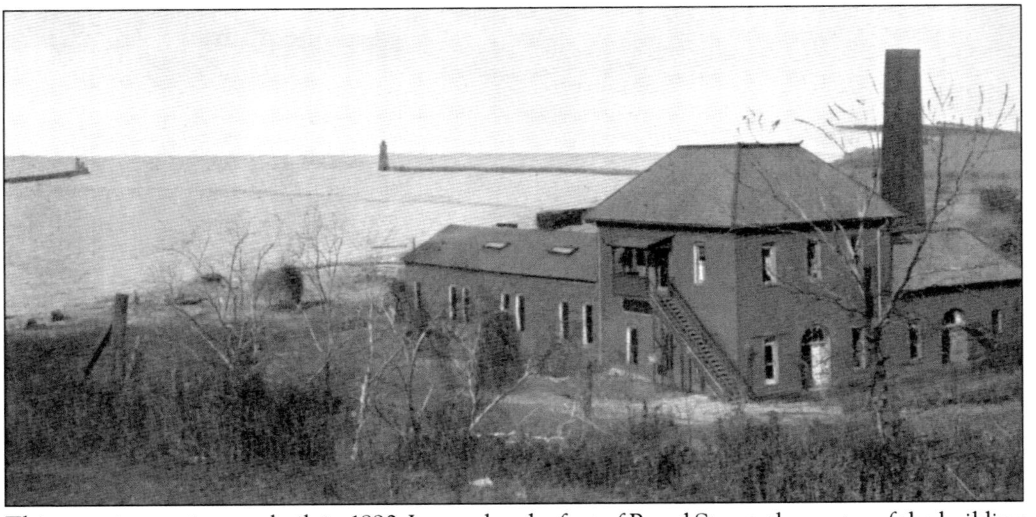

This pumping station was built in 1890. Located at the foot of Broad Street, the center of the building housed pumps capable of satisfying nearly four times the 1.25-million-gallon daily consumption. The left wing housed filtration tanks and the right wing coal-fired boilers. The water was pumped 2 miles, through a 12-inch pipe, to a 300,000-gallon iron standpipe. The standpipe was located on Liberty Avenue adjacent to the city cemetery near Wrights Avenue. (Mary Jane Fritz.)

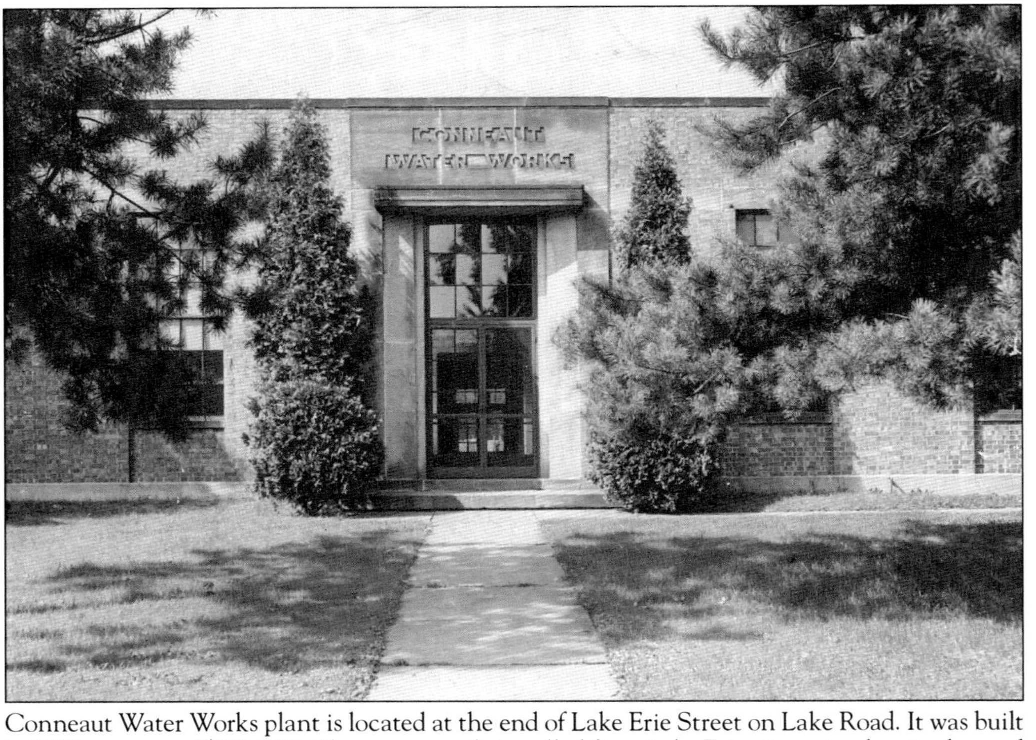

Conneaut Water Works plant is located at the end of Lake Erie Street on Lake Road. It was built in 1933 at a cost of $275,000. The water supply is pulled from Lake Erie via an intake pipe located behind the plant. The treatment capacity is 3 million gallons per day (mgd). In 1958, a 7.5 mgd wastewater treatment plant was built at 1206 Broad Street Extension for $1 million. (Conneaut Water Department.)

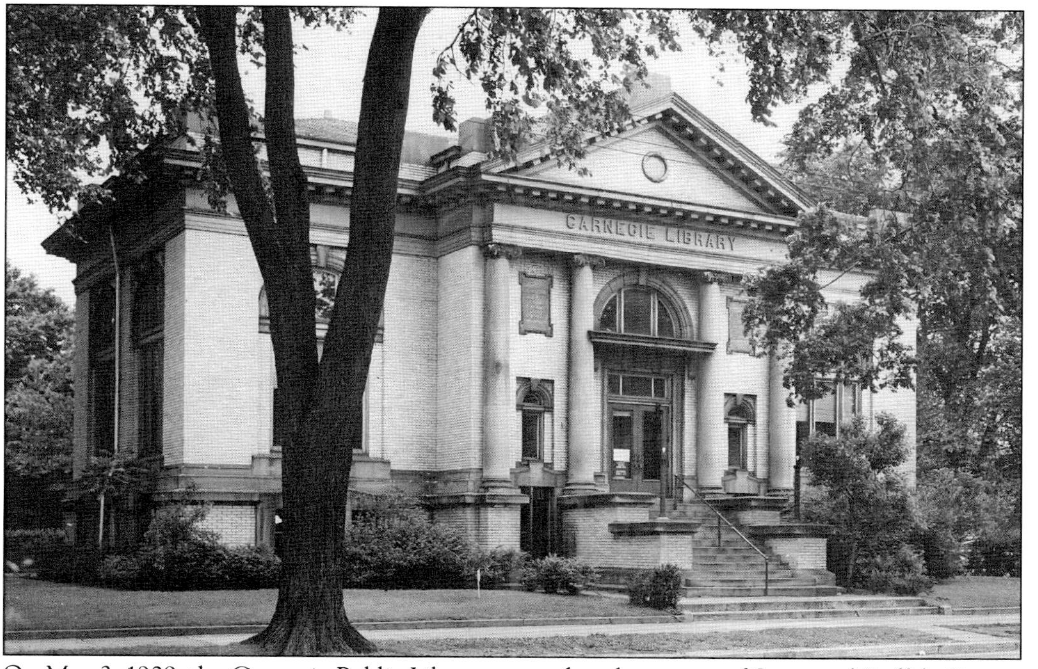

On May 3, 1909, the Carnegie Public Library opened at the corner of State and Buffalo Streets with 12,000 books. Conneaut's first lending library was the People's Free Library Association. The public was able to borrow free reading materials from its collection at H. H. Timby's bookstore. The library was built using funds from a 1905 library tax levy plus a 1907 Andrew Carnegie grant for $25,000.

To the left is the 1905 pressroom of *The Conneaut Daily Post-Herald*, which was created by the merger of the *Conneaut Evening Post* and the *Conneaut Herald* in 1896. In 1900, Conneaut's oldest newspaper, the *Conneaut Reporter* was also acquired. Both the *Post-Herald* and the *Reporter* were owned by John P. Rieg and Sagito J. Smith. The *Conneaut Reporter* was a direct descendent of the *Salem Advertiser*, which dated back to 1831.

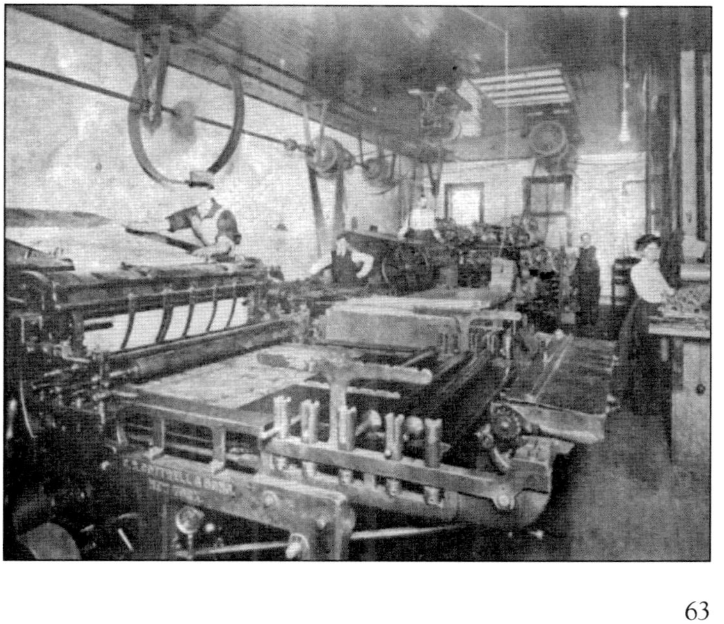

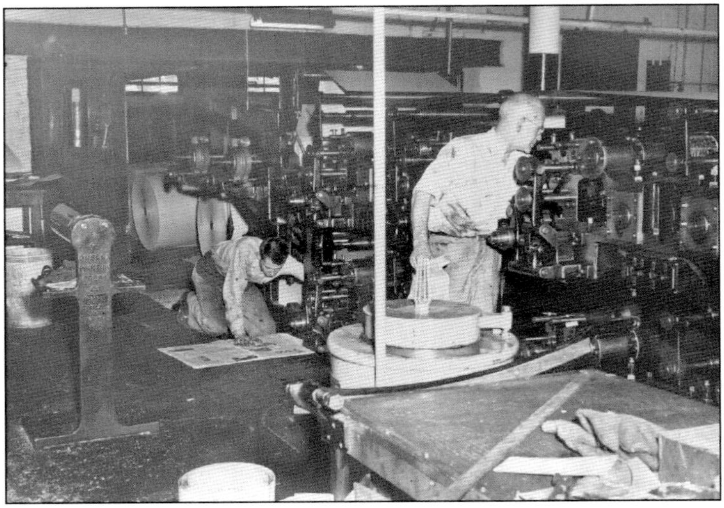

In September 1953, Maurice Hewitt (kneeling) and "Shorty" Lesperance ready the press for the daily *Conneaut News-Herald* edition. The press was capable of simultaneously printing 16 pages at 32,000 pages per hour. The *News-Herald* employed 30 employees with a distribution to 5,100 homes in northeast Ohio and northwest Pennsylvania.

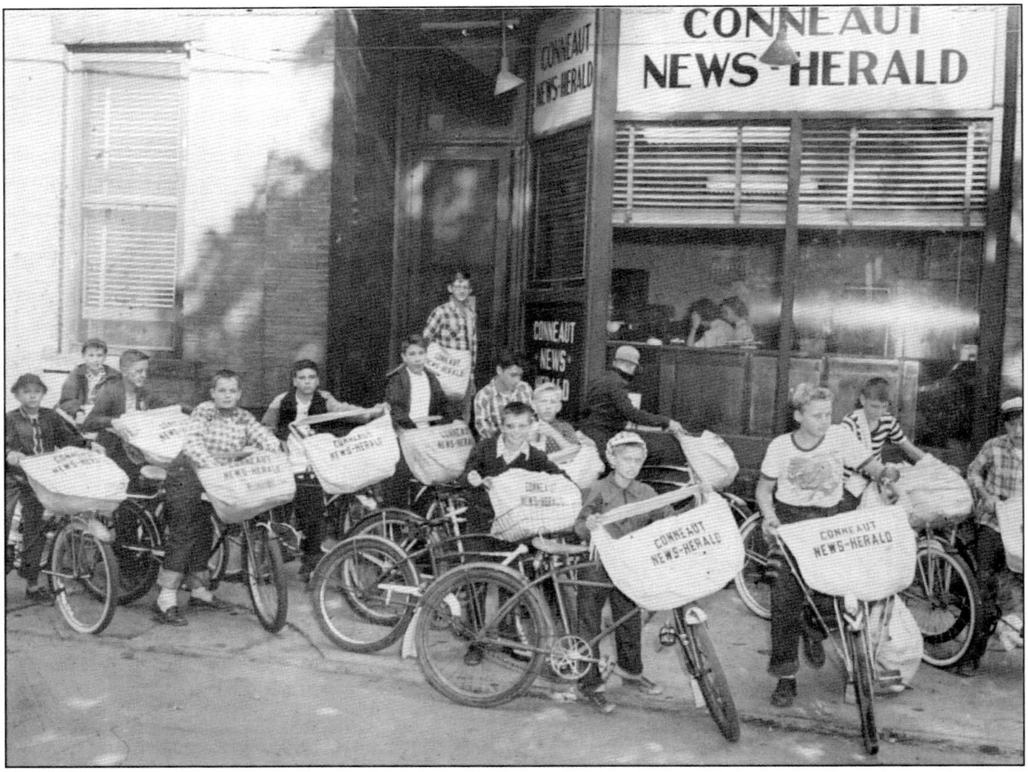

In September 1953, *Conneaut News-Herald* routes were coveted by young boys, because the routes were small and lucrative due to the large circulation. The *News-Herald* was created on January 1, 1907, by the merger of the *The Conneaut Daily Post-Herald*, owned by Rieg and Smith, and the *Conneaut Evening News*, owned by Purdy E. Bissell. Bissell established the *Evening Post*, which was Conneaut's first daily newspaper in 1892.

Five

Uptown

Conneaut's central business district has historically been known as "uptown." In Conneaut, this is where most of the professionals, such as attorneys, doctors, and dentists, had their offices in the early years. This is also where the government offices, utility companies, hotels, major retail stores, newspapers, and fraternal organizations were located. People from East Conneaut also referred to that part of town, on the other side of Conneaut Creek, as "crosstown."

On the following two pages, there is an uptown photograph from the late 1940s. In this chapter, the reader will take a trolley ride through uptown, beginning at city hall and moving east on Main Street, north on Harbor Street, and west on State Street. The photographs appear in chronological order, which provides a better prospective of how this part of town has changed over approximately 100 years.

At the end of the chapter, festivals and parades will be explored.

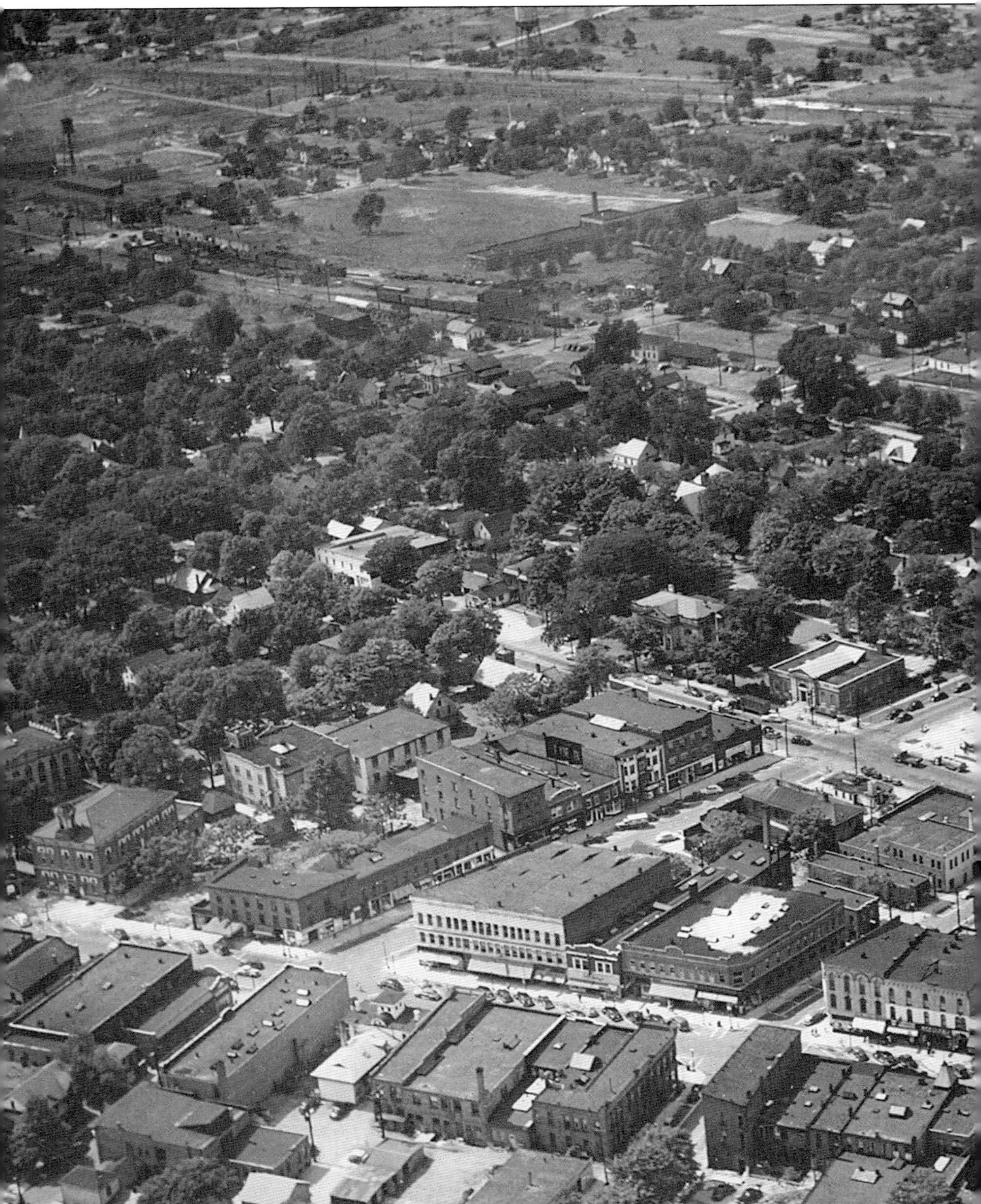

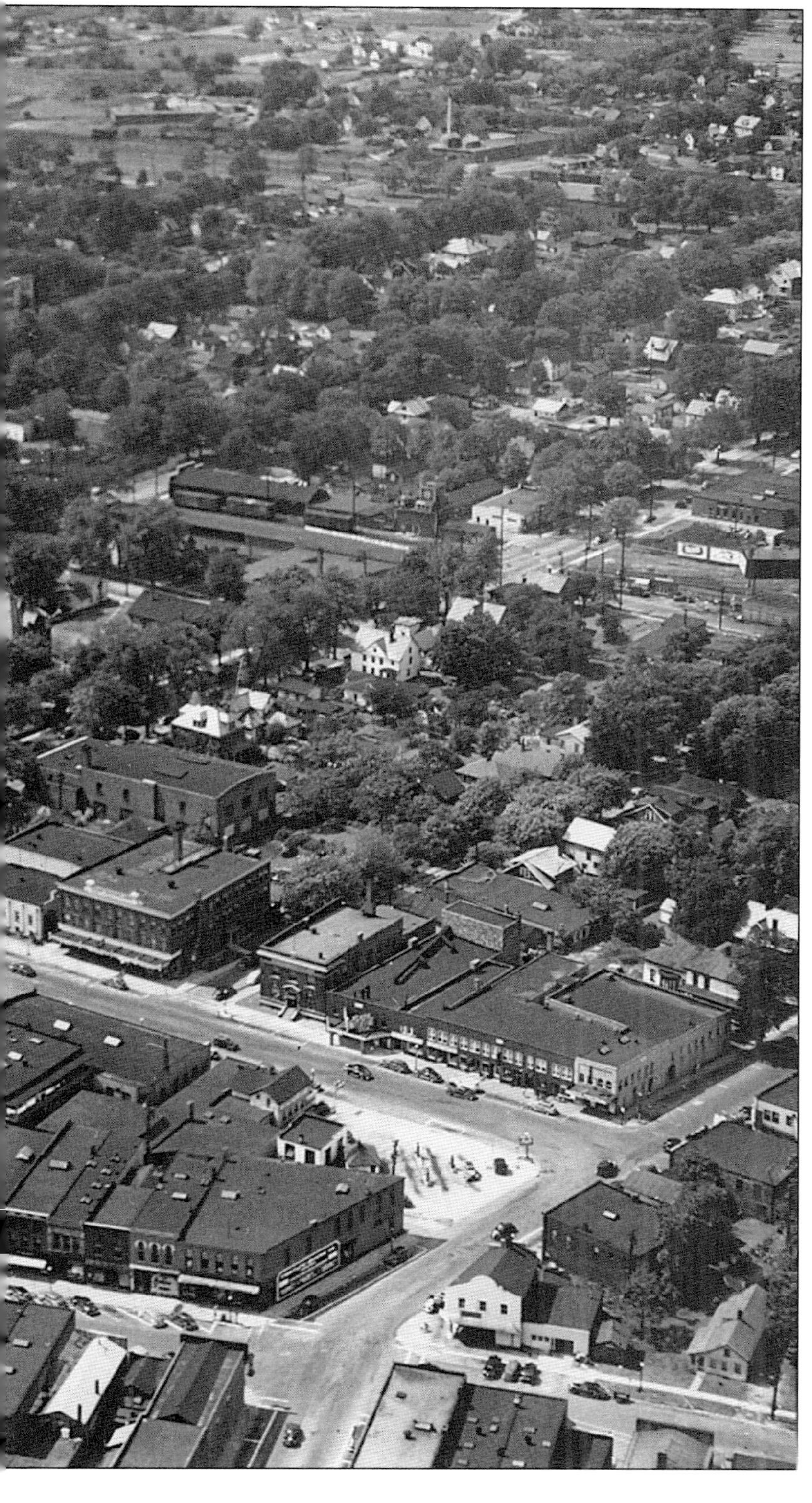

On page 66, the street running diagonally across the bottom of the photograph is Main Street. From left (west) to right (east) are Buffalo Street, city hall, Dorman block, Broad Street, Harwood block, Stanley block, and Washington Street. At the north end of Broad Street is the intersection with State Street. From east to west are the post office, Carnegie Library, and Ellsasser Garage (1887 electric plant). The long building in the upper right corner is the junior high school. On page 67, the street intersecting with Main Street at the bottom of the page is Harbor Street. On the left side is Fortune and Byers Hardware and on the right is the former Pennsylvania and Ohio Railway (trolley) terminal. From east to west at the State Street intersection are the creamery, Harbor Street, Ohio Theater, State Theater, The Conneaut Telephone Company, Washington Street, Cleveland Hotel, and the Masonic Temple.

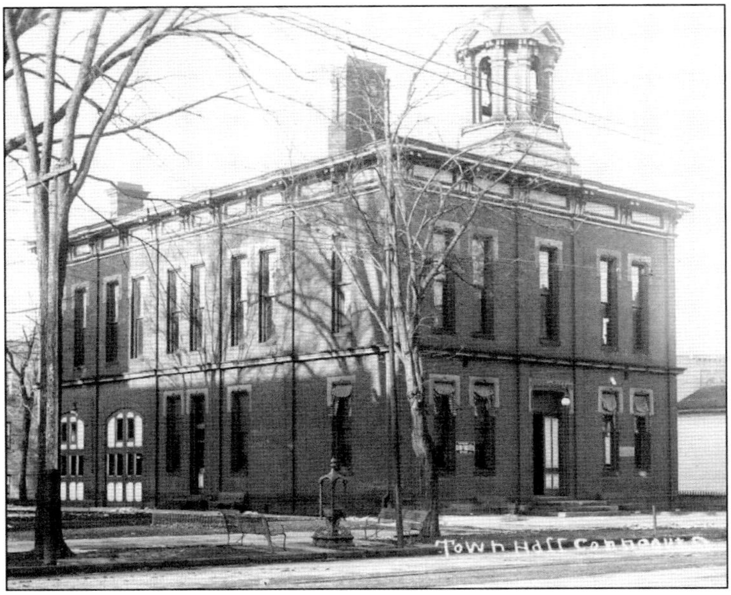

City hall was built on the northeast corner of Main and Buffalo Streets in 1876. When built, the first floor had offices, a voting room, fire pumper storage, and lockup. The horses' stables were in the Commercial House livery on Broad Street. An auditorium on the second floor was used for meetings, entertainment, and high school commencement exercises. The bell tower was used to notify volunteer firemen.

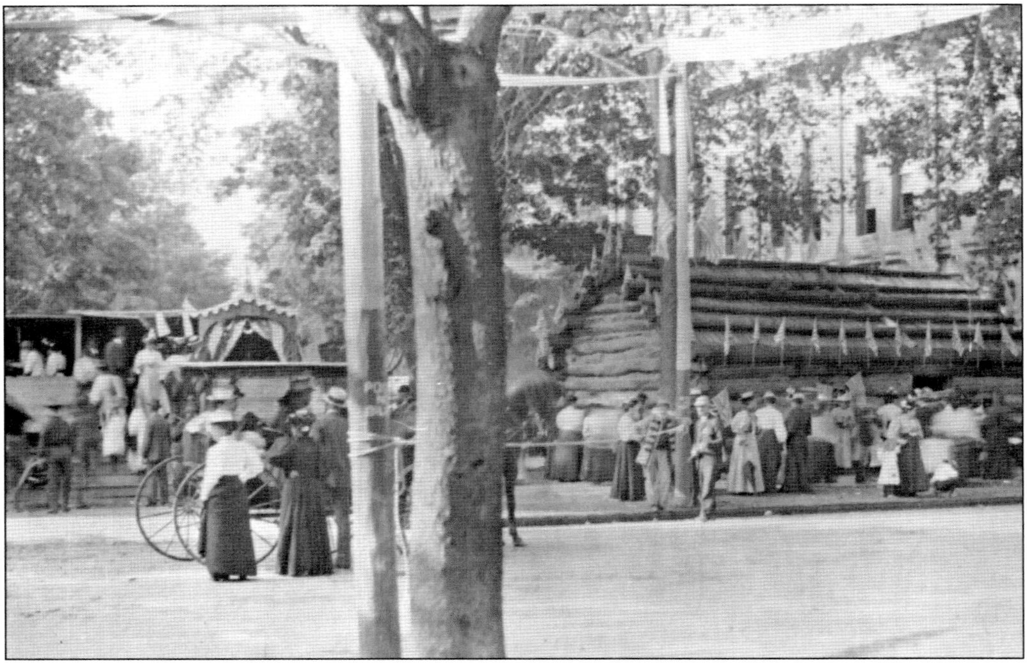

For Conneaut's centennial celebration on July 4, 1897, a replica of Cleaveland's Western Reserve surveying party headquarters was built on the west side of the town hall. The surveying party called the log storehouse "Stowe's Castle" in mock tribute to Joshua Stowe. As their commissary, Stowe was in charge of the food supplies. There was also a parade of ox carts, covered wagons, and people dressed in 1787 attire.

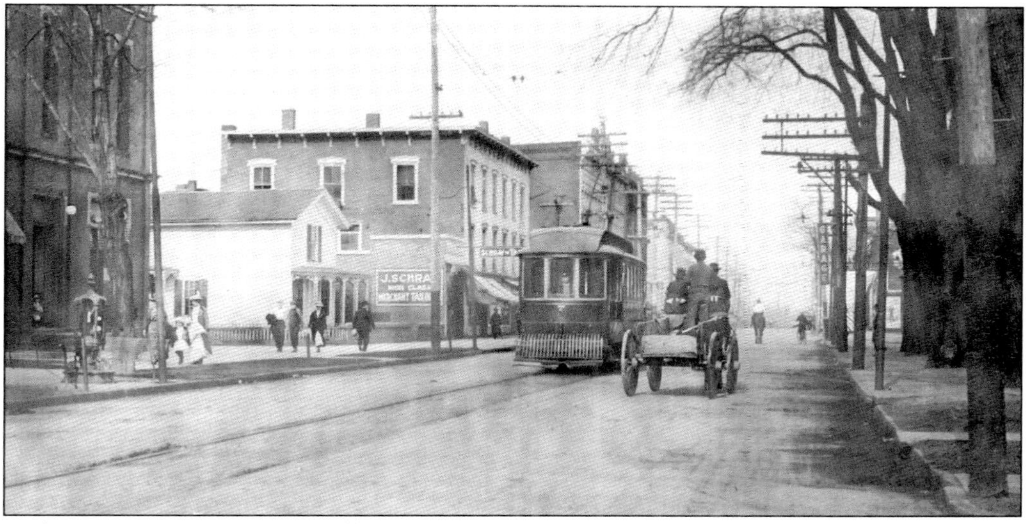

In this 1914 street scene, city trolley No. 7 is passing by city hall towards the west end. Conneaut had five city line trolleys, numbered 7 through 11. The large trees across the street from city hall are in the front yard of a house. The house was replaced by the brick building on the southeast corner of Main and Buffalo Streets in the 1920s.

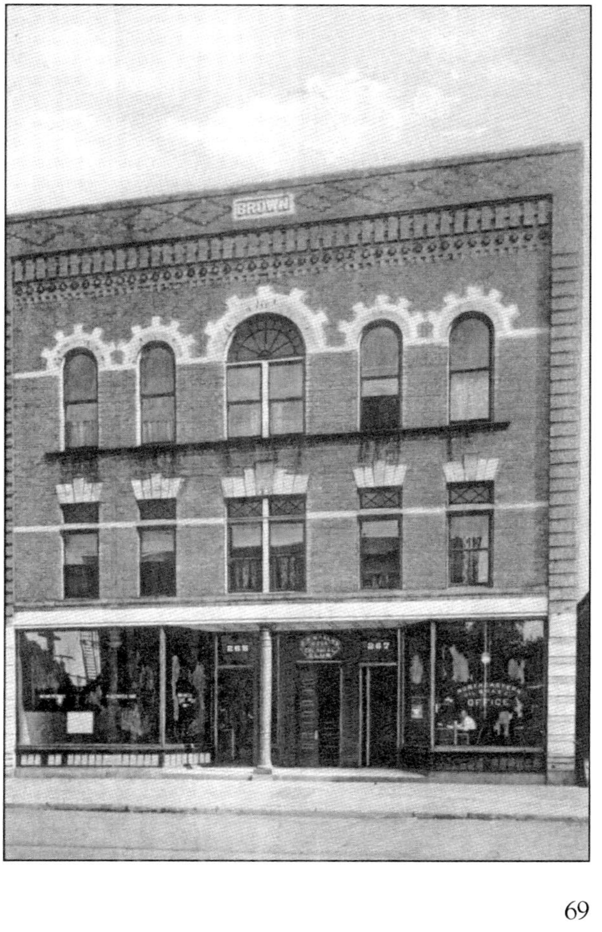

The Brown Building on the south side of Main Street was home to a movie theater at 265 (left) and at 267 sat the Northeastern Oil and Gas Company office; the Colonial Club was on the second floor and the Benevolent and Protective Order of Elks Lodge 256 was on the third floor. The most recognizable movie would be Shakespeare's *A Midsummer Night's Dream* starring Charlie Chaplin. (Mary Jane Fritts.)

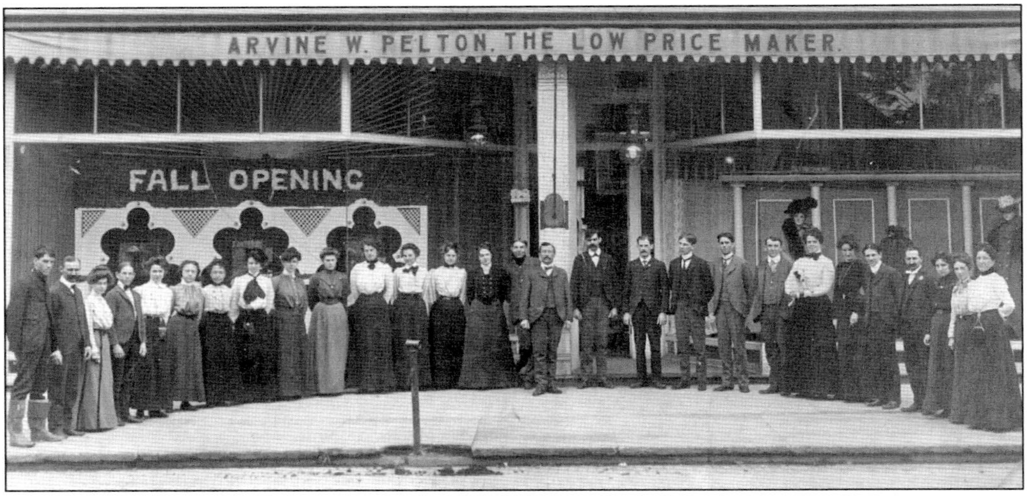

Arvine W. Pelton's department store featured dry goods, ladies clothing, millinery, curtains, carpet, and bedding. Bliss Harrington Business College was located in the floor above Pelton's prior to the offices in the photograph. In later years, the store's clientele and product lines would expand, requiring the store to occupy the first two floors of the Harwood block at Main and Broad Streets. Weldon Opera House was on the third floor.

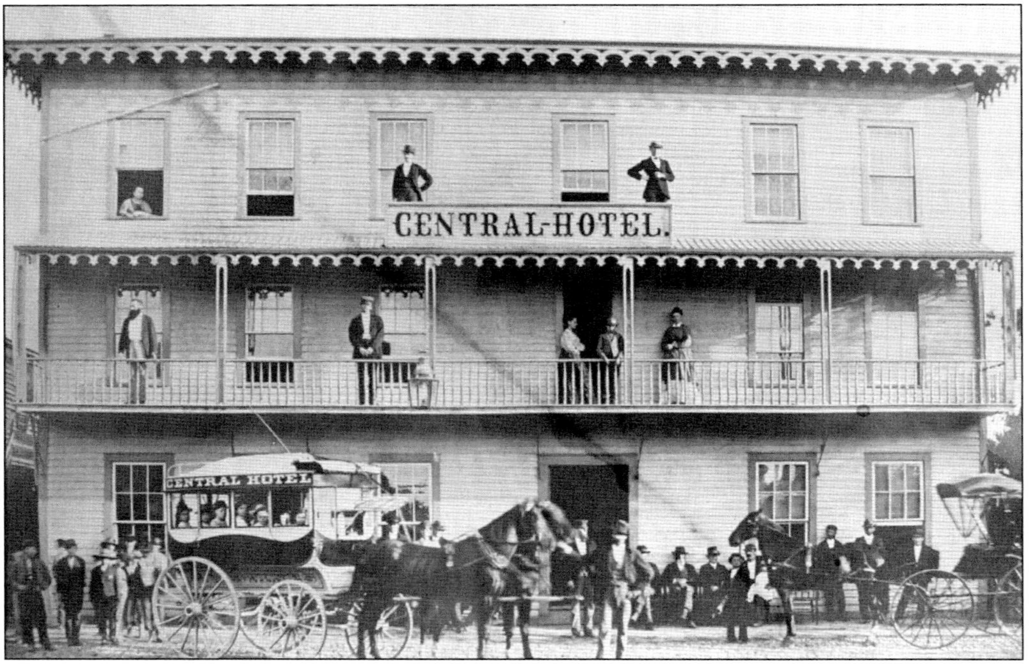

In this 1868 photograph, George Washington Stines, the owner of the Central Hotel, is in the center of the first-floor doorway. His wife, Esther, and son Frederick are standing in the doorway on the second-floor porch. His brother Burton is holding the horse's reins. The hotel was located at 241 Main Street.

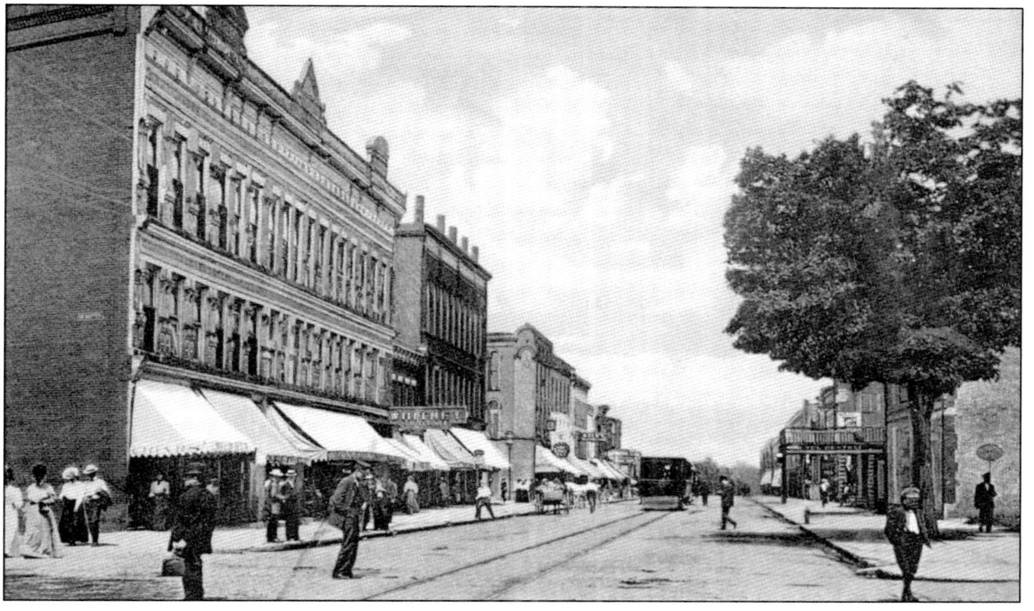

This postcard from around 1910 (above) looking east down Main Street from Broad Street has many versions. In the original photograph, utility poles line both sides of the street. The first building on the left is the Harwood block with various shops on the first floor, offices on the second, and the Weldon Opera House on the third. Frederick A. Mason, a streetcar conductor, would later operate the Mason Dance Academy on the third floor. The next three-story building is the Stanley block, home of Yeager's Dry Goods and the Conneaut Mutual Loan and Trust main office. The balcony over the sidewalk on the right is the Central Hotel. In this 1920s postcard (below), the Stanley block (left) has been rebuilt due to a fire. The Academy Theater (right) is now Gerdes Pharmacy.

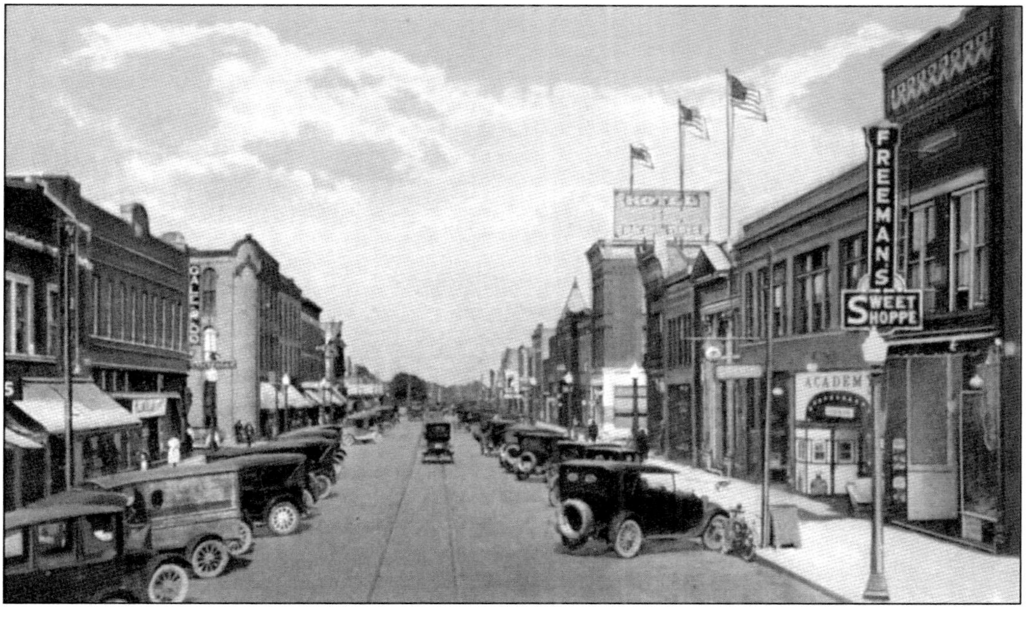

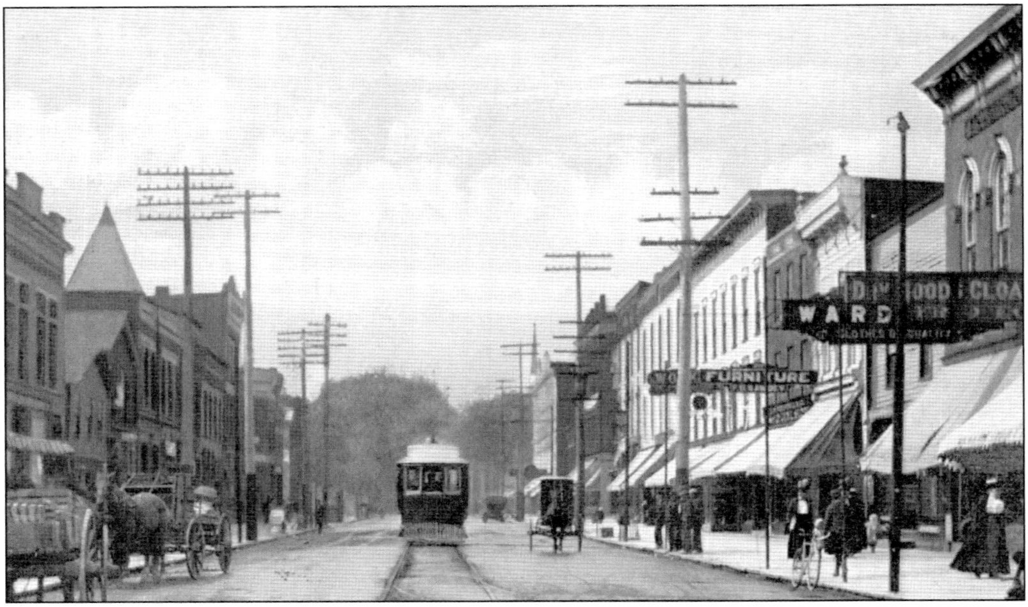

The above postcard from around 1910, looking west, provides an actual view of the utility poles on Main Street. On the left, 215 and 217, a barbershop and a bowling alley, are the last two frame buildings on the south side of Main Street. On the right, the Knights of Pythias, Grand Army of the Republic, Masonic Temple, and Knights of Columbus halls are located on the third floors from the Stanley block (five chimneys) to the Cleveland block (light colored building behind the "Furniture" sign). The steeple at the far right is the Congregational church. In the 1920s postcard below, the old Congregational church was razed in 1915 to be replaced by the May Record-Findley Memorial Chapel Annex. A drugstore, at 208, is the last frame building on Main Street.

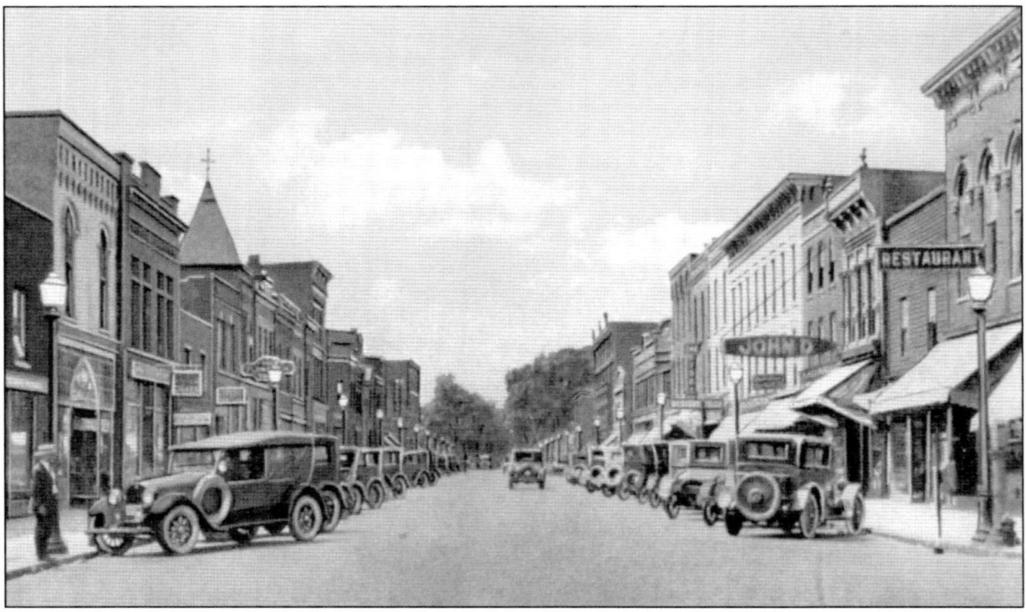

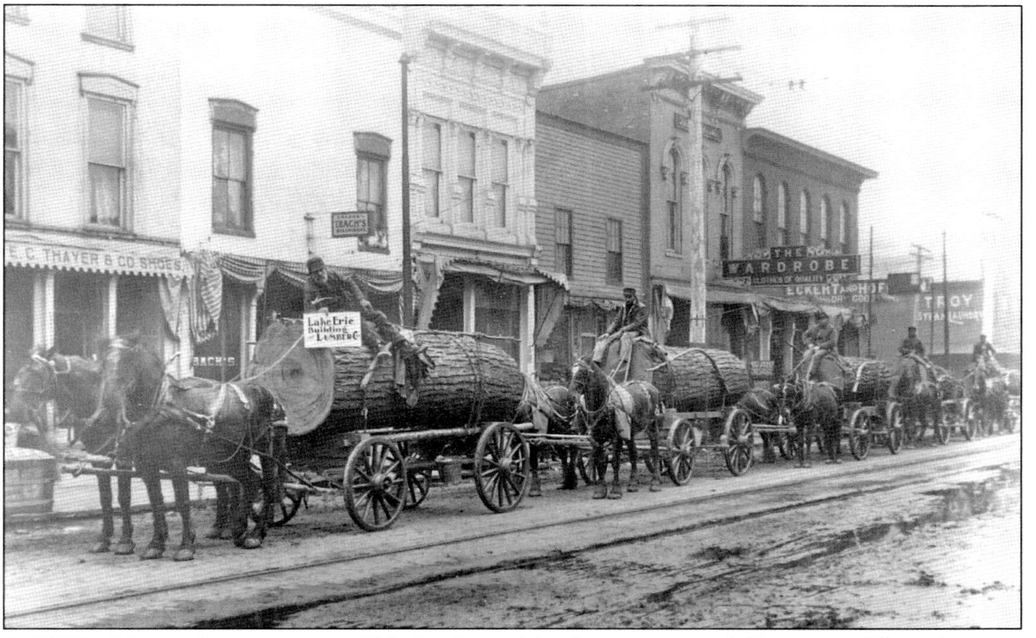

In 1902, old growth trees were still being logged in the area around Conneaut. These logs are headed west along Main Street. The Troy Steam Laundry (right) was on the northeast corner of Main and Harbor Streets. These logs were most likely destined for Thompson Brothers Saw and Planning Mill at 650 Buffalo Street, where they produced lumber from logs. Thompson Brothers later moved west of Whitney Road.

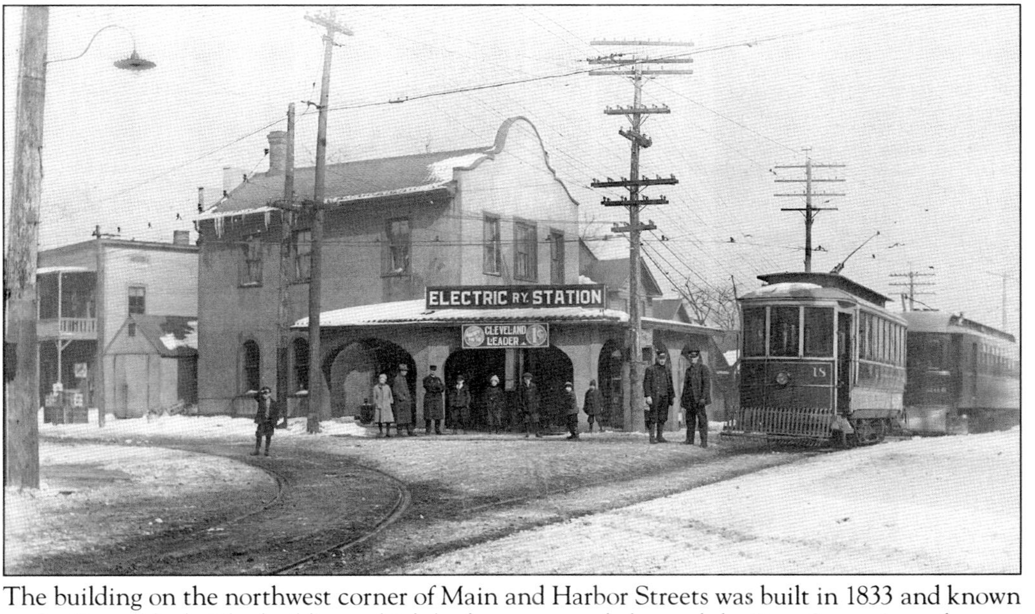

The building on the northwest corner of Main and Harbor Streets was built in 1833 and known as the Keyes block. The building, which had most recently housed the Troy Steam Laundry, was remodeled and became Conneaut's trolley station. This was the point for transfers and ticket purchases for the Pennsylvania and Ohio Railway city line to the harbor, Pennsylvania and Ohio interurban line to Ashtabula and Jefferson, or Conneaut and Erie Railway line to Erie.

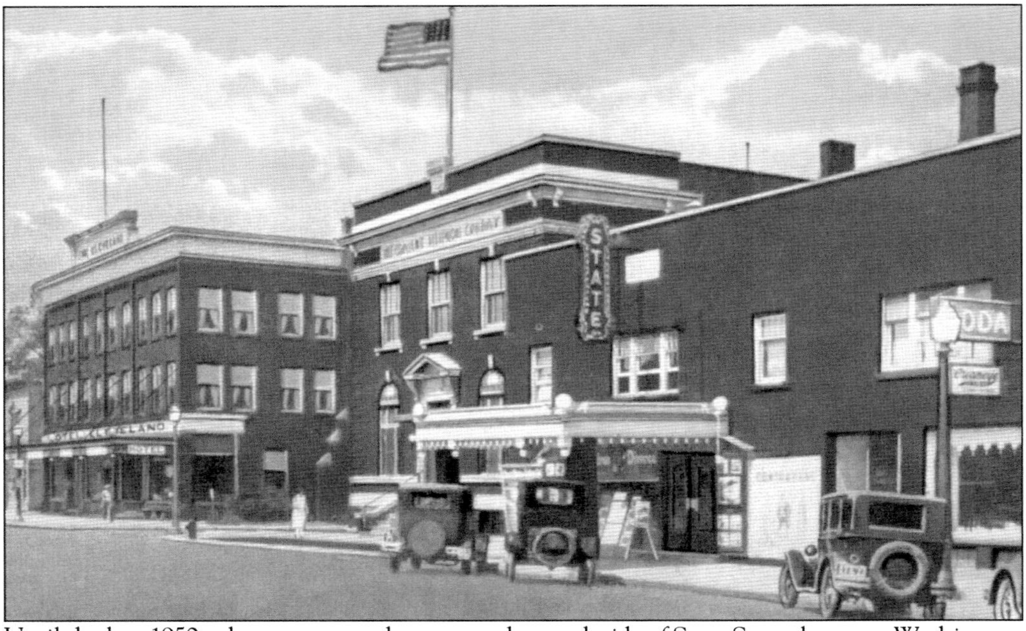

Until the late 1950s, there were two theaters on the north side of State Street between Washington and Harbor Streets. The entrance to the Ohio Theater was on the corner of State and Harbor Streets. It was torn down around 1960 and replaced with a drive-in restaurant. The State Theater closed a few years later. Pictured below is the interior of the State Theater. There was also a balcony with seats on both sides of the projection booth. A few years earlier, the Republic Theater, located at 320 State Street, featured a balcony and box seats on both sides of the screen.

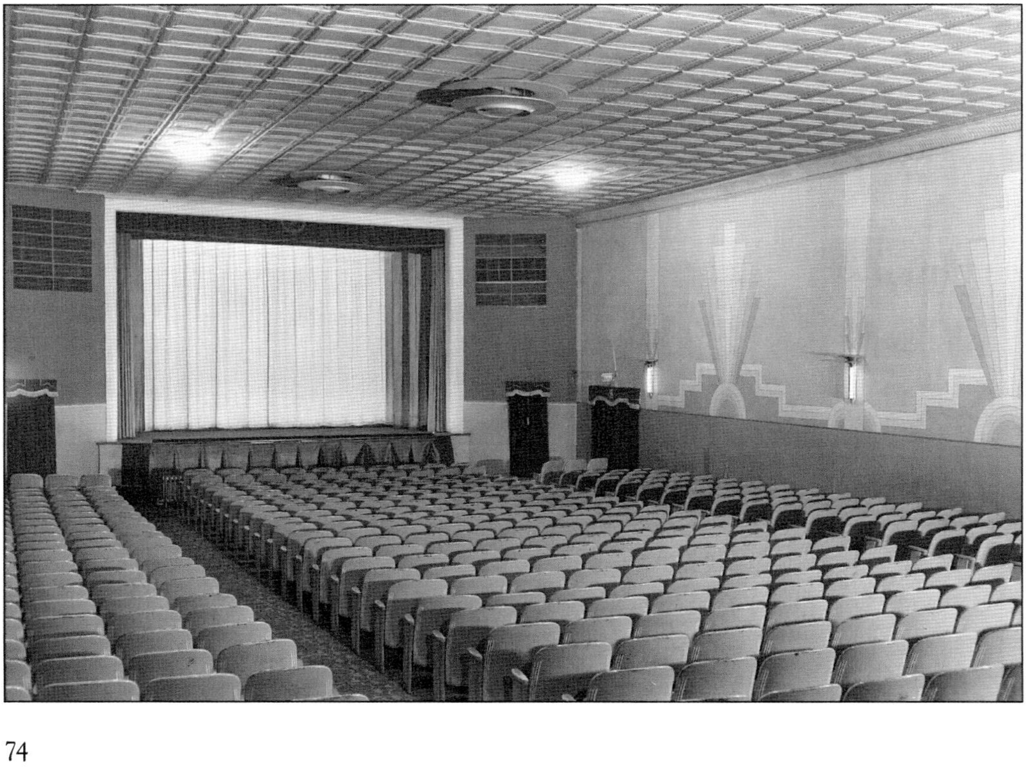

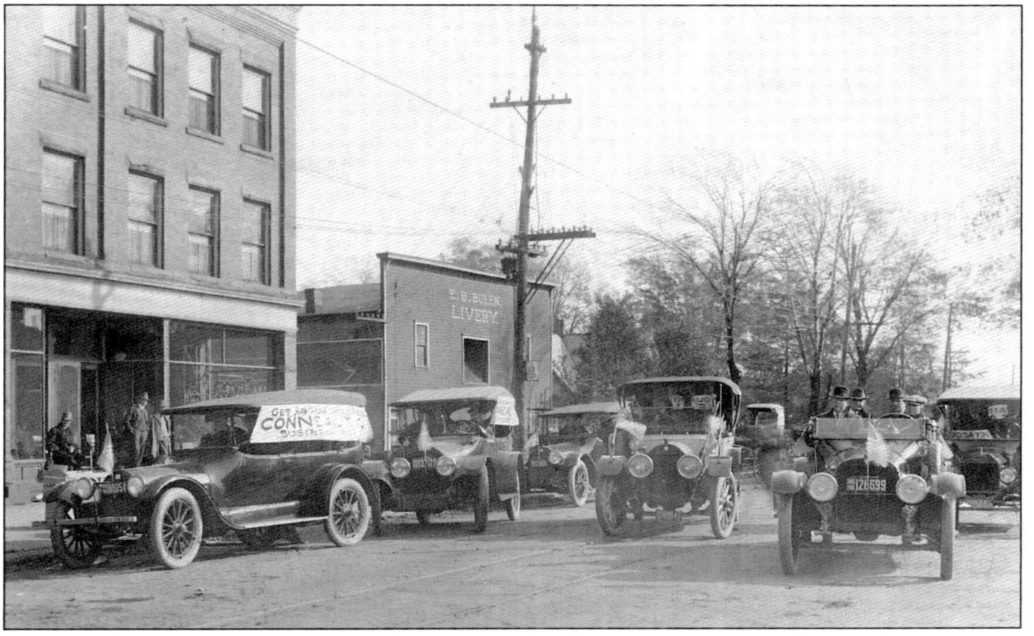

In 1916, the Conneaut Chamber of Commerce held a "Get Acquainted Tour" for Conneaut businessmen. The purpose of the tour was for chamber members to get acquainted with each other by providing the opportunity to showcase facilities, products, and services. They traveled by caravan from the chamber of commerce office in the Cleveland Hotel (left) throughout the city of Conneaut.

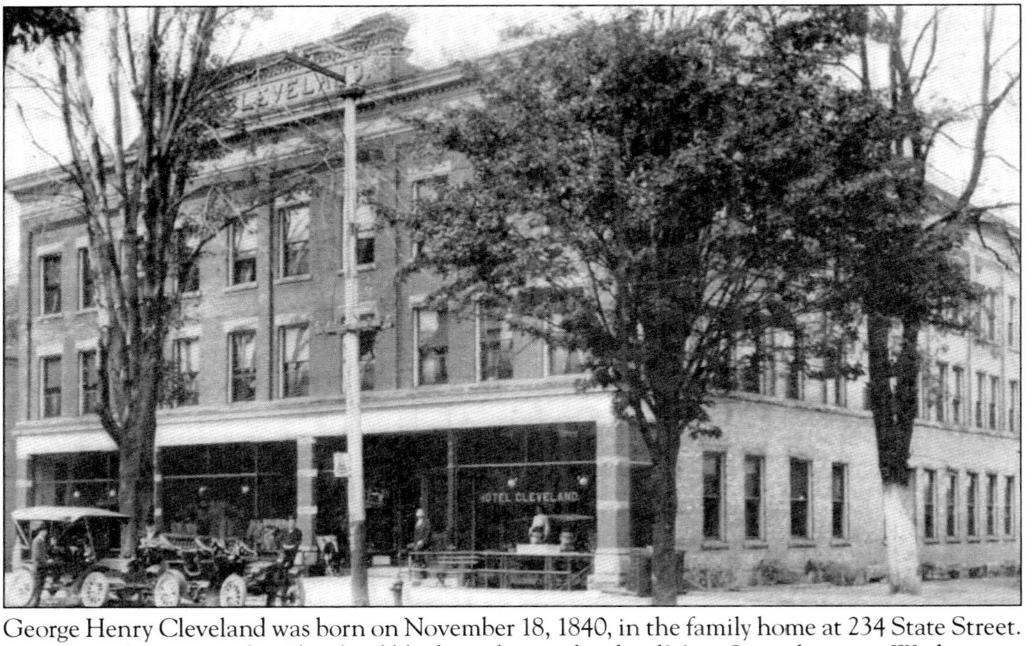

George Henry Cleveland was born on November 18, 1840, in the family home at 234 State Street. He operated a store in the Cleveland block on the north side of Main Street between Washington and Harbor Streets from 1862 to 1888. In 1903, Cleveland's sons, Merritt and Clarence, built the Cleveland Hotel on the site of the old family home. (Mary Jane Fritts.)

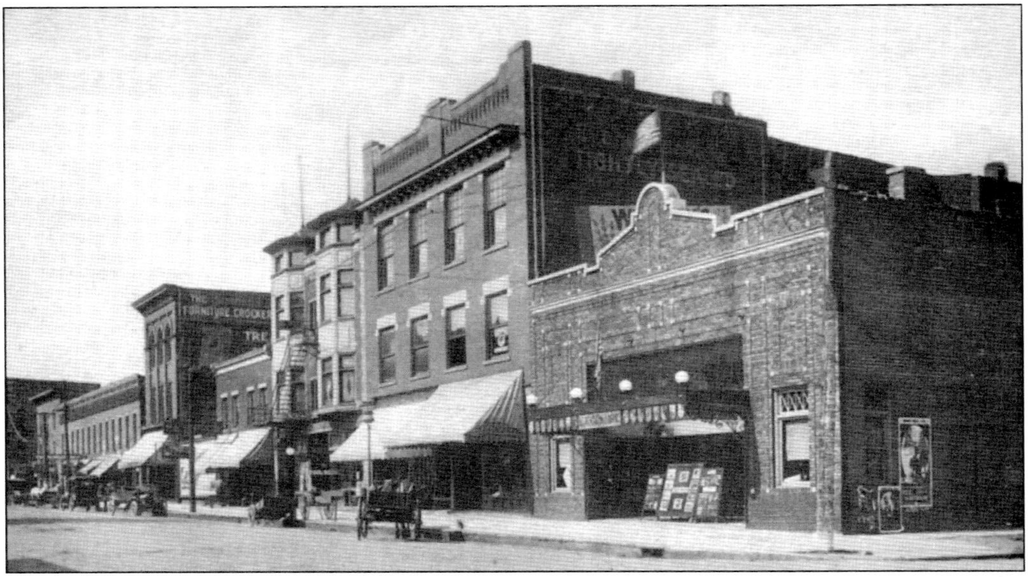

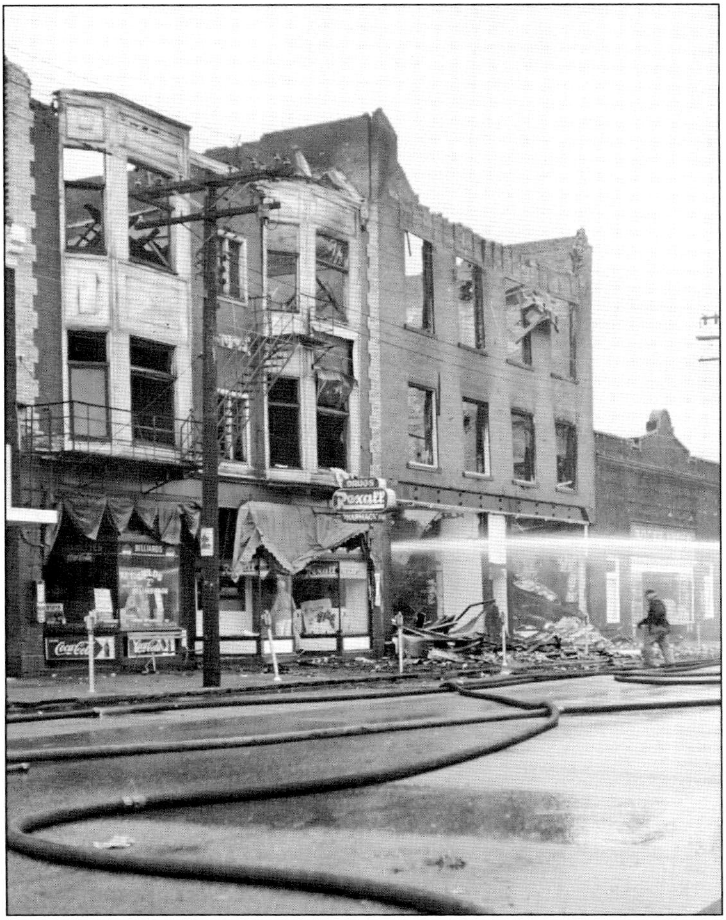

In this *c.* 1920 postcard (above), the LaGrande Theater shows moving pictures on the southwest corner of Broad and Main Streets. In the next building south on Broad Street, there is a grocer and a store selling paint and wallpaper. The next building, with double bay windows, was the Culp block. In the Culp block, there was a variety store and pool hall. In the 1950s, there was a fire (left), which destroyed the two buildings south of the former LaGrande Theater. The Culp block was rebuilt as a nondescript yellow brick building with Childs Pool Hall relocating to the second floor and Perkoski's Pharmacy on the first floor. The other building became a parking lot. (Above, Mary Jane Fritts.)

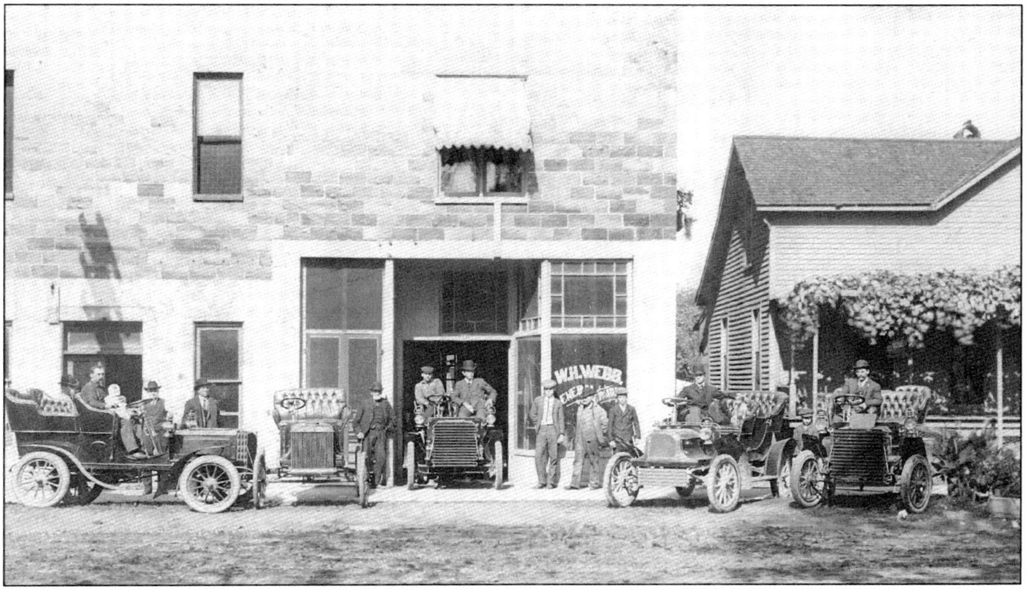

Upon relocation of the Conneaut Electric Light plant to the foot of Main Street hill, William Webb opened a Rambler dealership in his building. The 1904 Rambler on the left is sitting directly in front of Webb's original light plant. A second story has been added, and the building has been widened. The Rambler and other automobiles have unusually large radiators, which is an indication that Webb had converted them to "steamers."

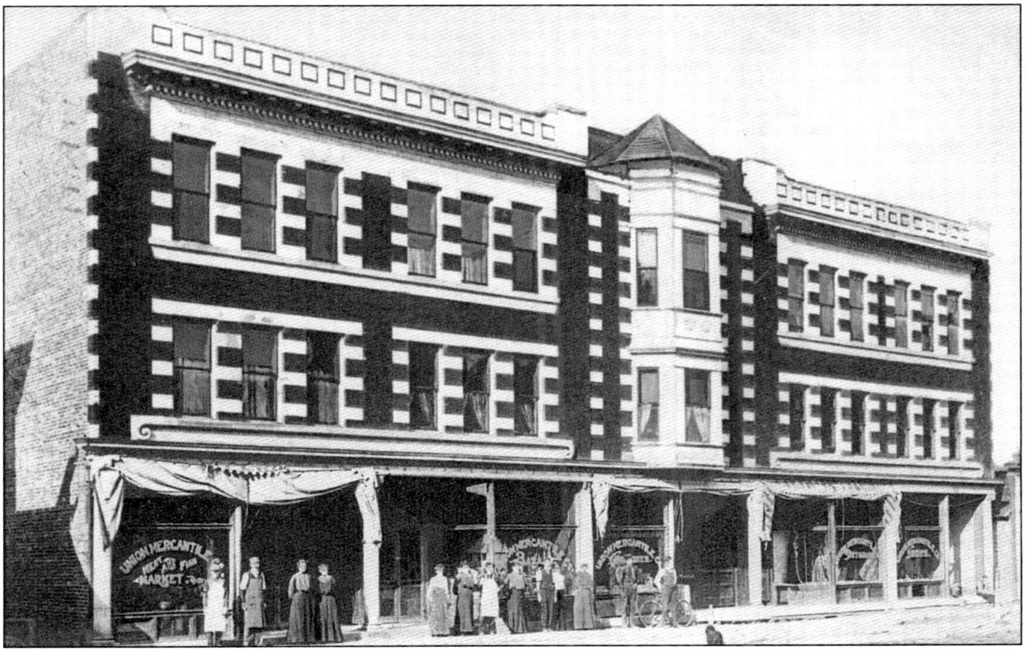

The Harrington block, at 450–458 State Street, was built around 1900. In this *c.* 1910 postcard, it was occupied by the Union Mercantile Company, which was the most diversified retail store in Conneaut's history. Union Mercantile sold meat, fish, sundries, groceries, dry goods, and shoes. (Mary Jane Fritts.)

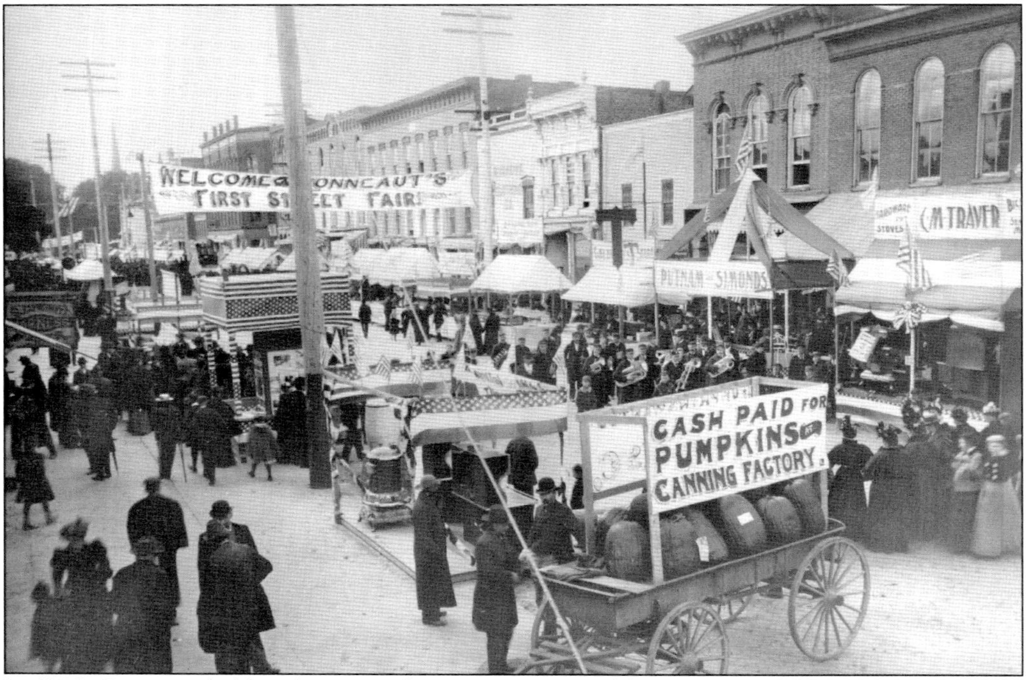

In 1896, Conneaut held its first street fair on Main Street (above), where merchants set up booths on the street and bands played for the entertainment of the townspeople. The instrument under the "I" in the Putnam and Simonds furniture store sign is a double-belled euphonium. The instrument was introduced in 1888 and is chiefly known as a collector's item or museum piece today. In 1957, the instrument regained its place in history through its mention in the song "76 Trombones" from Meredith Wilson's musical *The Music Man*. Around 1905, the street fair (below) was far more sophisticated, featuring carnival rides and curiosities such as Me-Mo, the alligator boy.

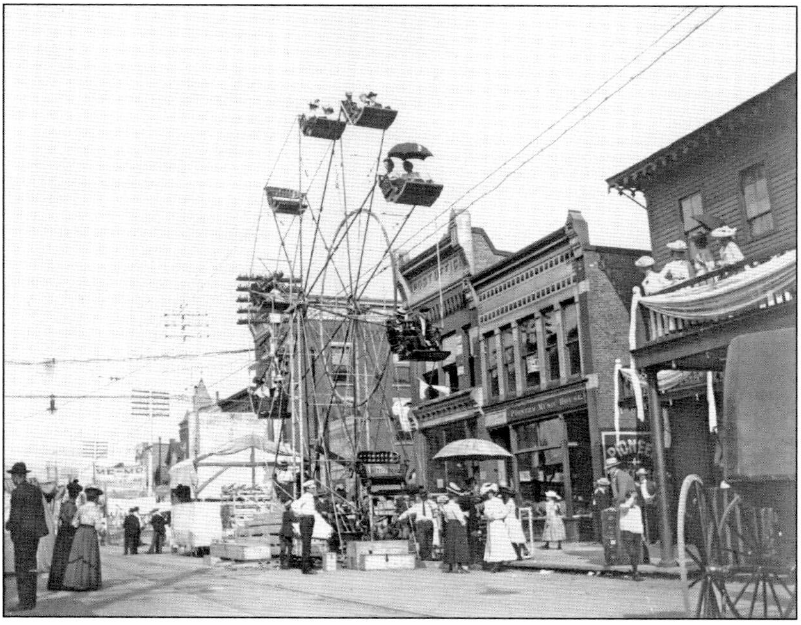

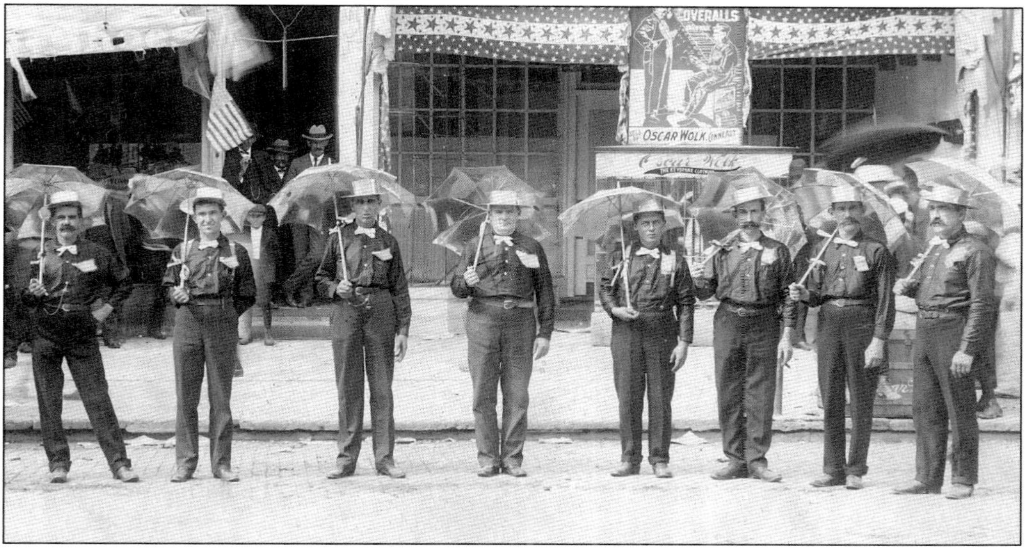

Just because people did not smile in photographs around 1905 does not mean they did not have a sense of humor. Perhaps it was their way of saying, "I'd rather be wearing Headlight Overalls from Oscar Wolk's." For whatever reason, these men, who were secure enough with their manhood to parade down Main Street carrying parasols, worked in the Nickel Plate tin shop.

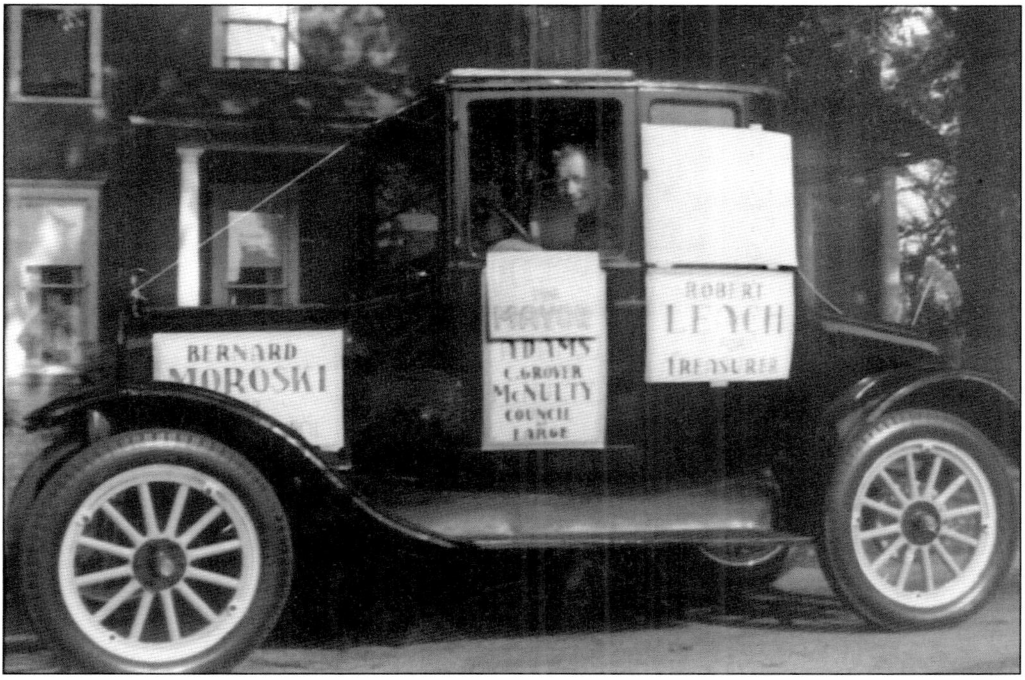

Tom Perkio drives his 1923 Model T Ford during his successful 1957 campaign for mayor of Conneaut. Perkio, a Democrat, passed out campaign cards, which had "GOP–Good Old Perkio" printed on them. This Model T has been in the Perkio family since it was purchased in 1923. The car will be on display in 2010 at the Conneaut Area Historical Society Museum. (Steve Perkio Jr.)

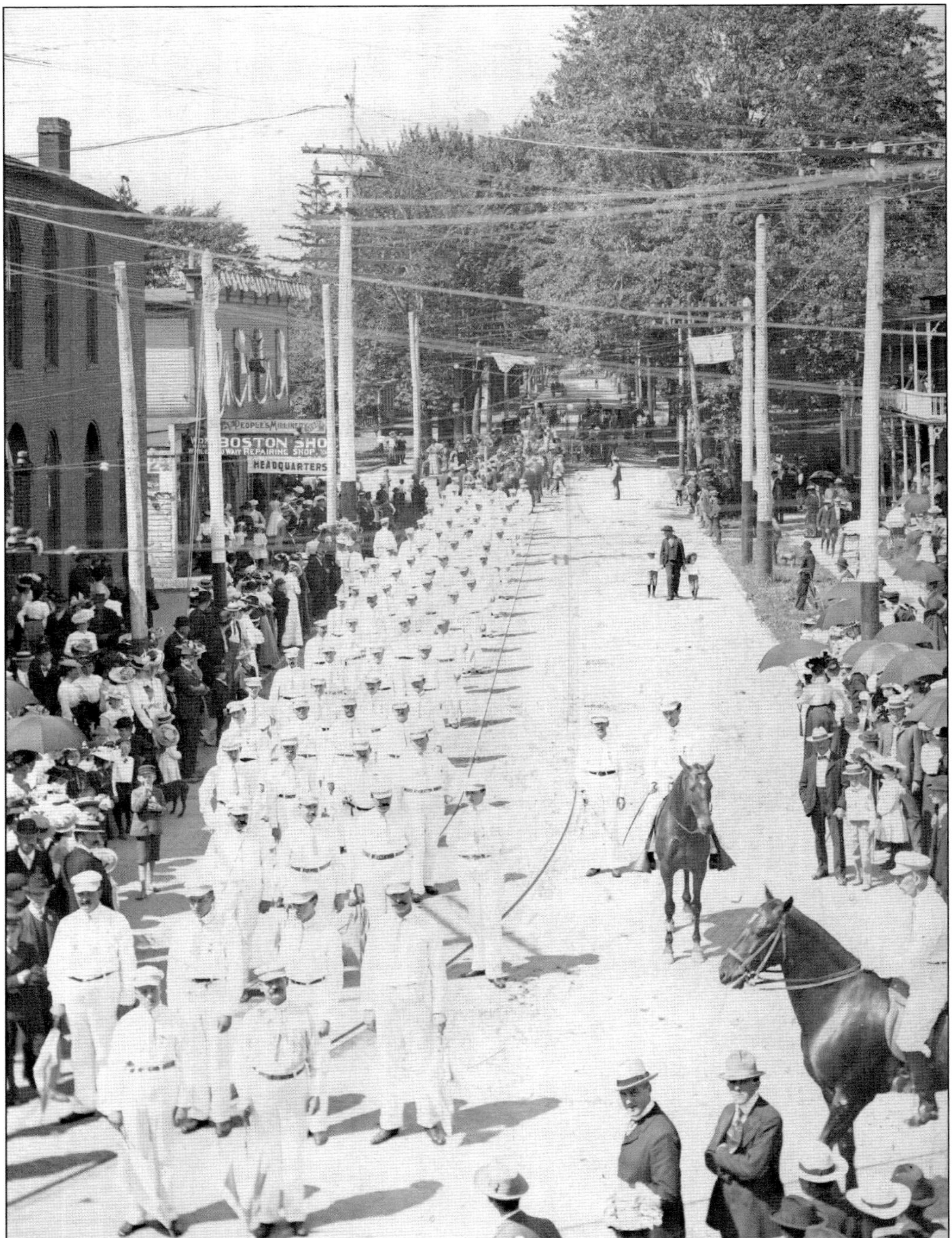

Around 1903, over 70 members of the Conneaut Elks Club dress in white and line up between Main and Washington Streets for a parade. The Elks Club headquarters occupies the second floor of the building on the southwest corner of State and Harbor Streets above the Peoples Millinery Store (left). The building on the right side of Harbor Street, with the second-floor porch, is the Wellington House.

Six

Around Town

This chapter will explore Conneaut's neighborhoods, unique individuals, their homes, and small businesses.

In many cities across America, the polar opposite of uptown is what is known as downtown—a less robust business district featuring an earthier element. Conneaut never had a downtown per se, but it had the harbor, which did meet the necessary criteria. Before automobiles became the prevalent mode of transportation, the harbor was home to small shops, a bank, post office, and most of the city's bars. In the 1920s, during Prohibition, it was also the realm of Conneaut's criminal element known as the Black Hand.

In the late 1920s, Clara D. Peet donated 75 acres of land as a campground for Conneaut's young people. Located on Conneaut Creek, in Amboy, it featured a dining hall, log cabins, and a dam in its heyday. After the buildings were gone, Camp Peet continued to be used by scouting groups.

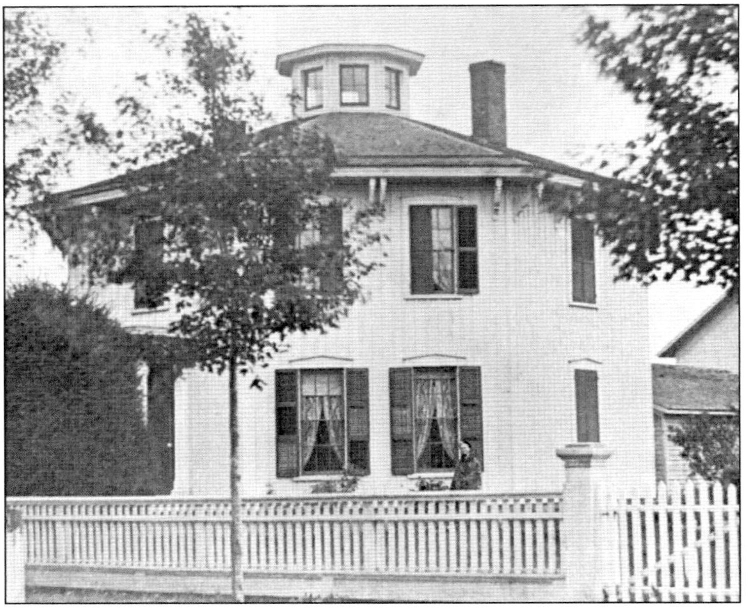

David Cummins built this octagon home at 310 Liberty Street in 1863. Cummins Canning Company was born in the kitchen where Cummins began by canning tomatoes. As an Underground Railroad station, it featured a trap door from the home's pantry to an underground tunnel extending 150 feet to Conneaut Creek. The house, now owned by the Conneaut Area Historical Society, is being restored as a museum.

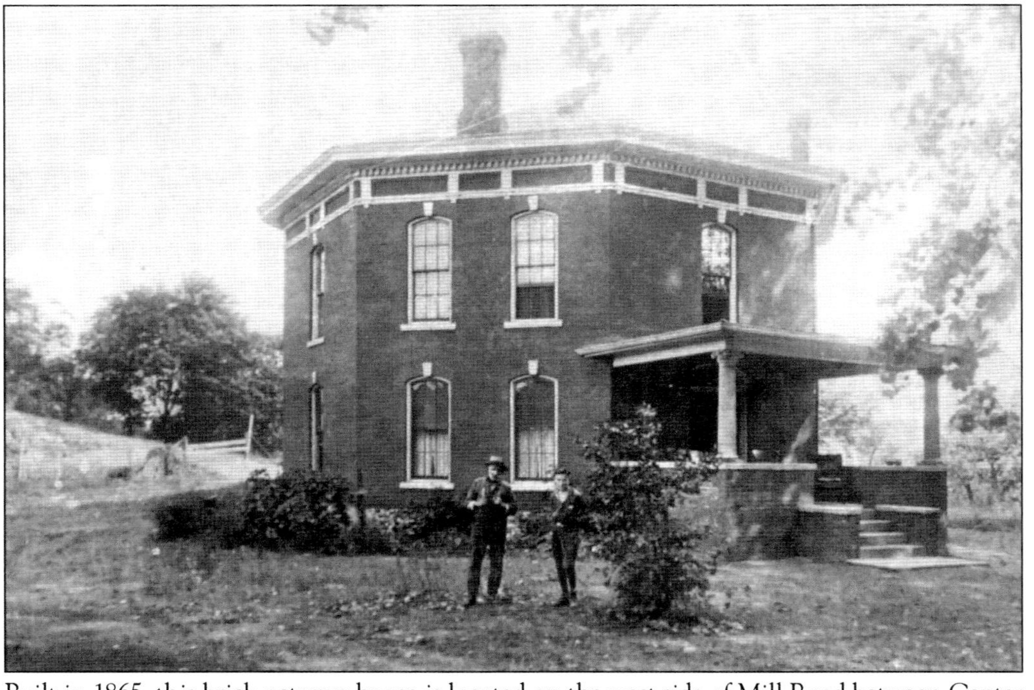

Built in 1865, this brick octagon house is located on the west side of Mill Road between Center Street and Welton Road. Downstairs, the living and dining rooms have square corners, but all other rooms feature wedge-shaped corners. There are five bedrooms upstairs, one bedroom downstairs, and two bathrooms. The basement walls are made of 24-inch-thick stone, and the house walls are double brick with an insulating air space between.

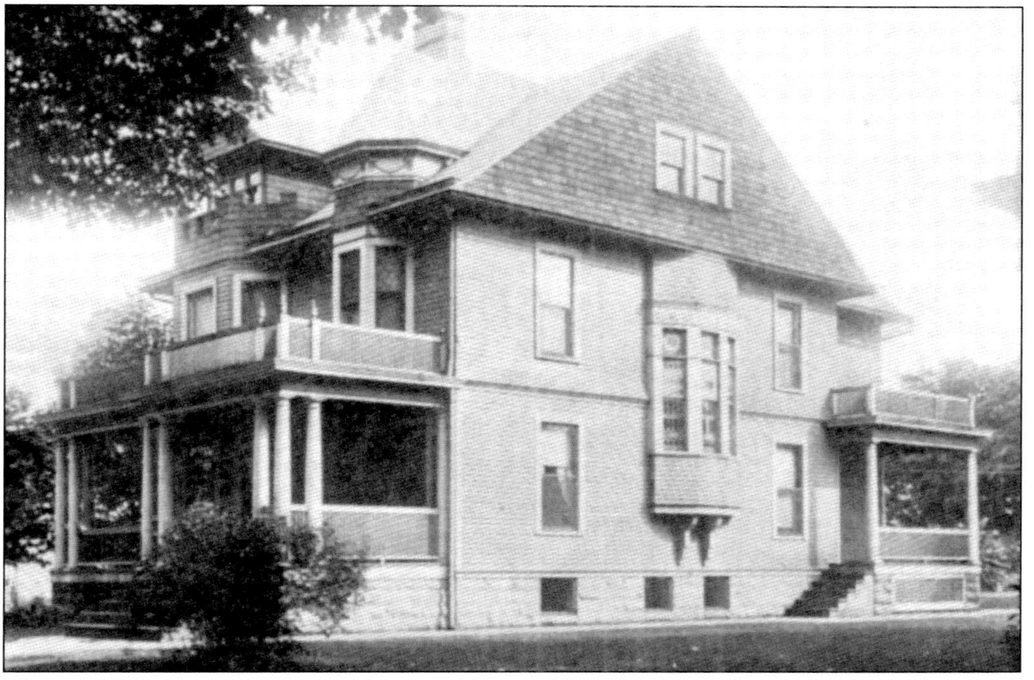

This is the Winthrop Cushman Pelton residence at 359 Main Street. Born in Kinsman, he joined the Ohio Volunteer Cavalry, fought, and was taken prisoner at Ream's Station, Virginia, in 1864. He came to Conneaut in 1882 and founded the C. W. Pelton and Brother store at the corner of Main and Washington Streets. He was also president of the Conneaut Mutual Loan Association and the Conneaut Electric Lighting Company.

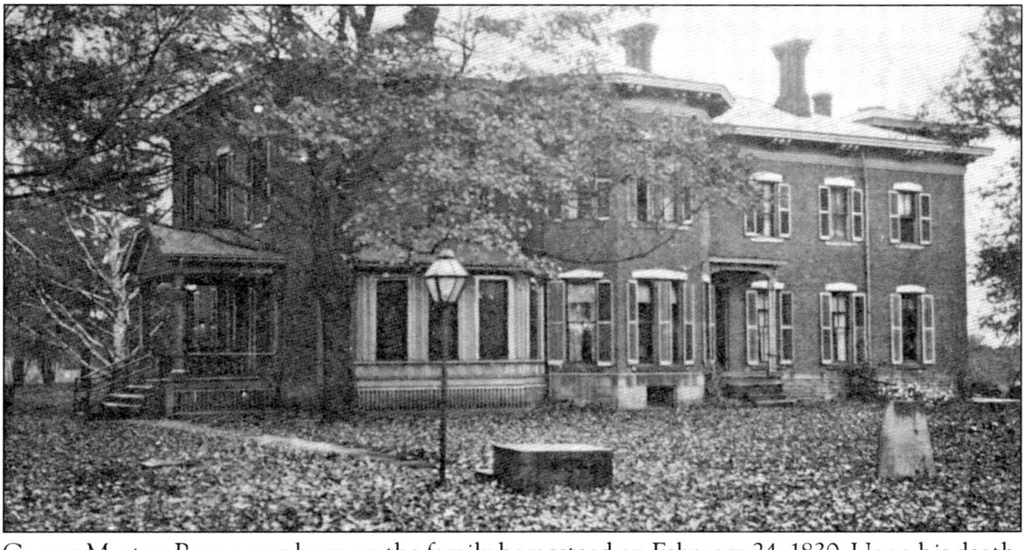

George Morton Brown was born on the family homestead on February 24, 1830. Upon his death, he bequeathed his family's estate, on the northwest corner of West Main Street and Brown Avenue, for use as Conneaut's hospital. Brown was educated in Conneaut schools and fought with the 15th Army during the Civil War. He was vice president of the Conneaut Mutual Loan and Trust Company and one of Conneaut's most influential citizens. (Mary Jane Fritts.)

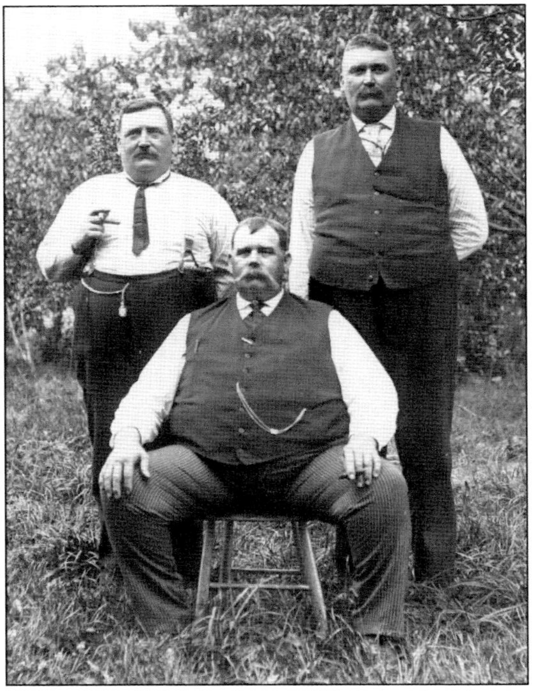

This photograph of (left to right) John Reddington, Clayton M. Webb, and John Laughlin was taken at the Brotherhood of Locomotive Engineers picnic on August 11, 1905, at Tarry-A-While on Lake Road. These Nickel Plate engineers were considered to be the largest in the country with a combined weight close to 1,000 pounds. (Joan Webb Barnett.)

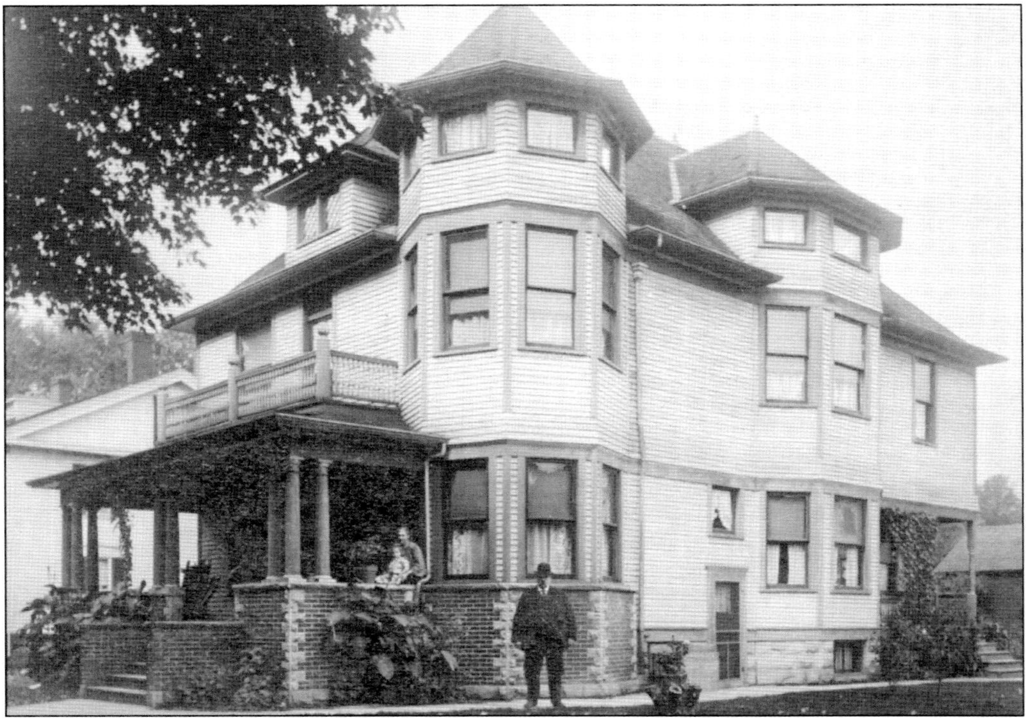

In 1900, locomotive engineer Clayton M. Webb was believed to be the largest man in Conneaut, when he built this house at 408 Main Street. It was said that Webb, at 400 pounds, was able to slip in and out of his locomotive cab as gracefully as any man half his size. (Joan Webb Barnett.)

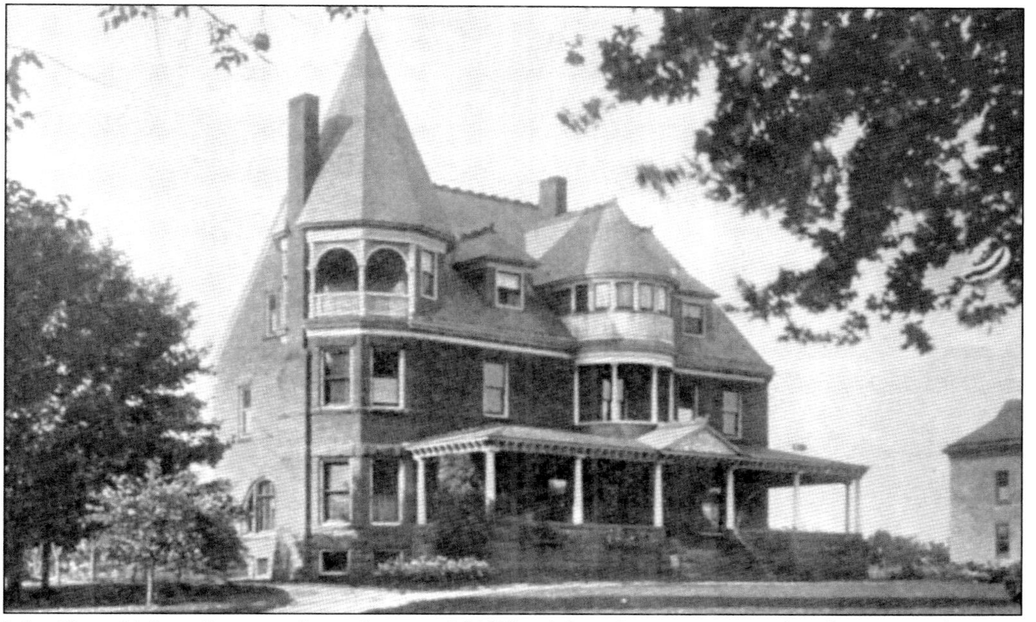

John D. and Mary Cummins's residence at 884 West Main Street was completed in 1897. The attic contains a 2,000-gallon copper tank to collect roof water, which is gravity fed throughout the home. John's primary occupation was with the family canning business. One of the organizers of the Conneaut Mutual Loan and Trust Company, he also served as vice president of the Pennsylvania and Ohio Railway Company interurban streetcar line.

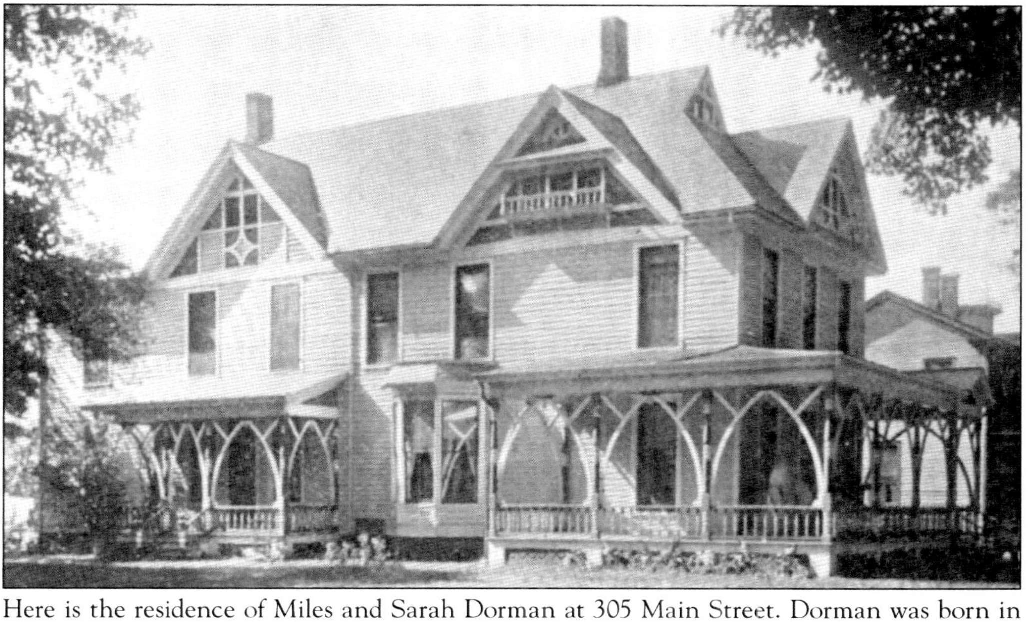

Here is the residence of Miles and Sarah Dorman at 305 Main Street. Dorman was born in Erie County, Pennsylvania, in 1837. After a successful career in livery and hotel businesses in Pennsylvania, he moved to Conneaut in 1871. At various times he owned the Central Hotel on Main Street, the Commercial House at Broad and Main Streets, and a livery business. He served on Conneaut City Council from 1878 to 1886.

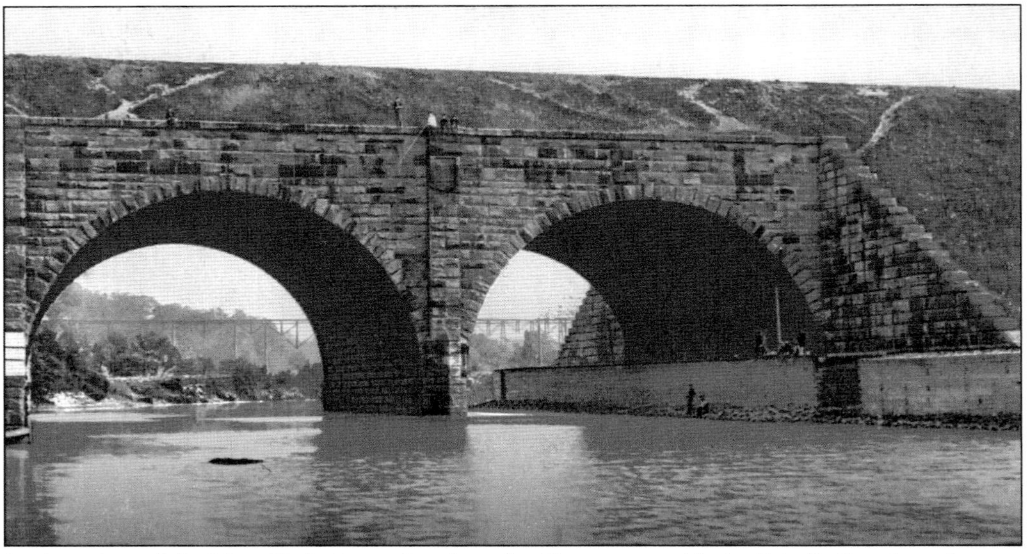

In this view of the New York Central arches, the Bessemer and Lake Erie Railroad tracks pass under the right arch and the Nickel Plate railroad trestle can be seen in the background. This was a favorite fishing hole for neighborhood boys. Skinny-dipping was also a favorite pastime on the other side of the left arch and in the old brick plant quarry beyond the right arch. (Mary Jane Fritts.)

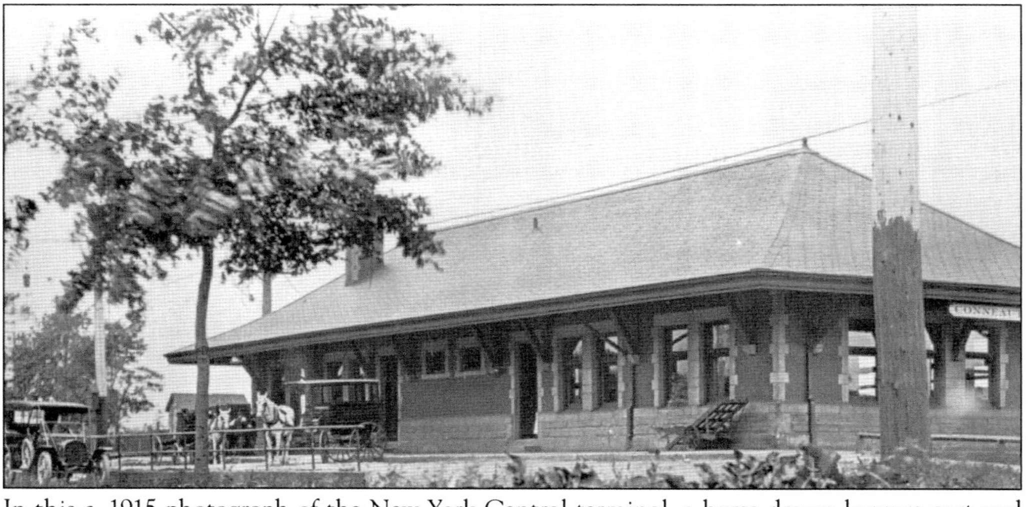

In this c. 1915 photograph of the New York Central terminal, a horse-drawn luggage cart and a passenger carriage await the arrival of a passenger train, while a Cadillac touring car waits on Depot Street. Passengers could also catch the city line trolley to take them north on Broad Street to the Lake View Park Hotel at the end of Park Avenue or south on Sandusky Street to the Cleveland Hotel on State Street.

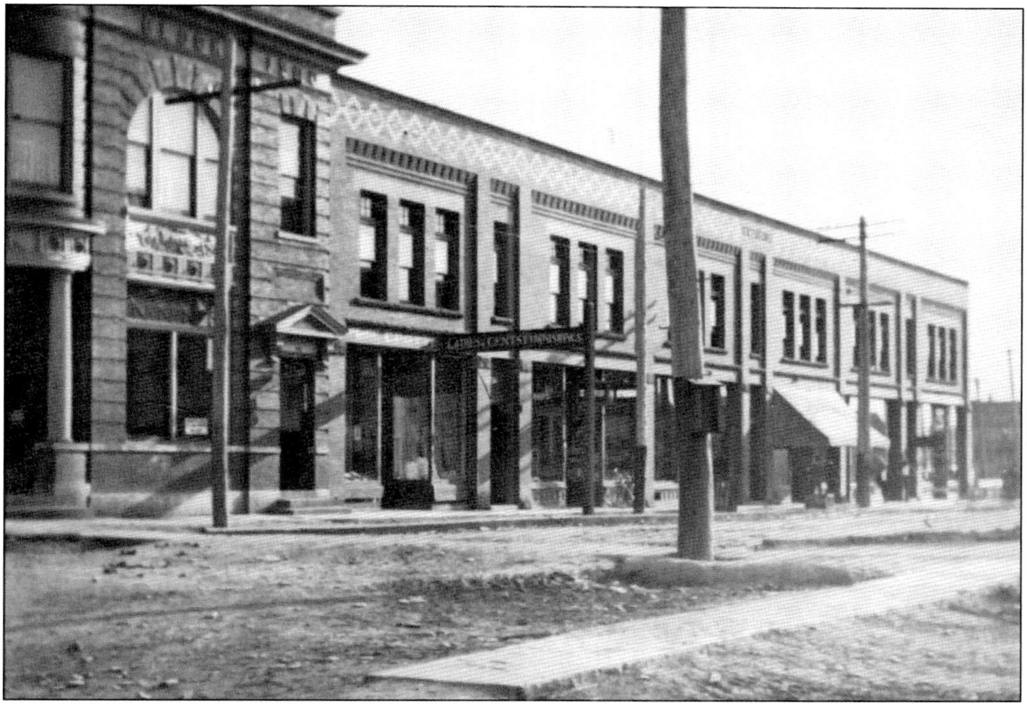

In 1900, the newly built Marine block (left), on the west side of Day Street, contained the Marine Bank, post office, shoe store, men's store, and barbershop. In the Mutual block, along the south side of Park Avenue, were two clothing stores, a drug store, grocery, bakery, and hardware store. To the right, the Finnish Temperance Hall can be seen before the addition of the cupola. On February 25, 1904, a fire destroyed both buildings (below), resulting in a $100,000 loss. Above the smoldering ruins, the steeple of the Finnish Lutheran Church (right) can be seen. (Above, Ashtabula County Historical Society.)

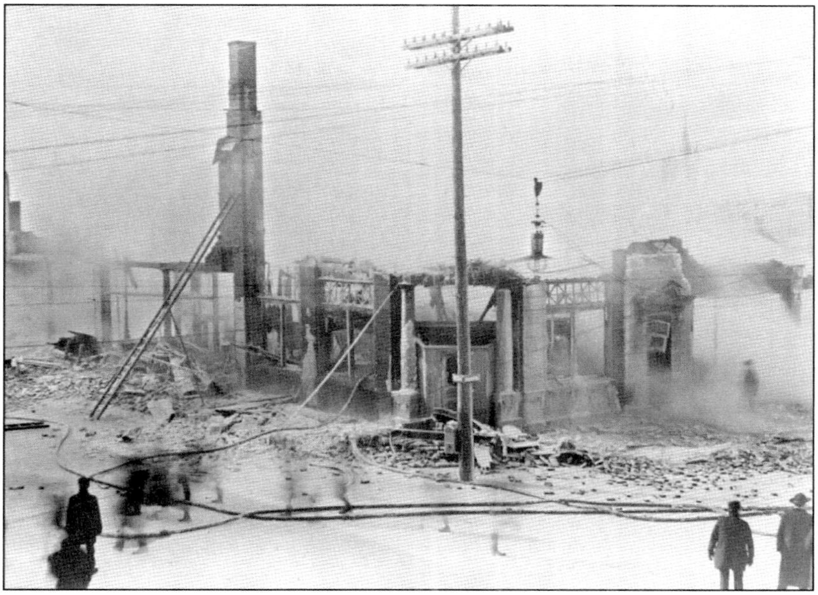

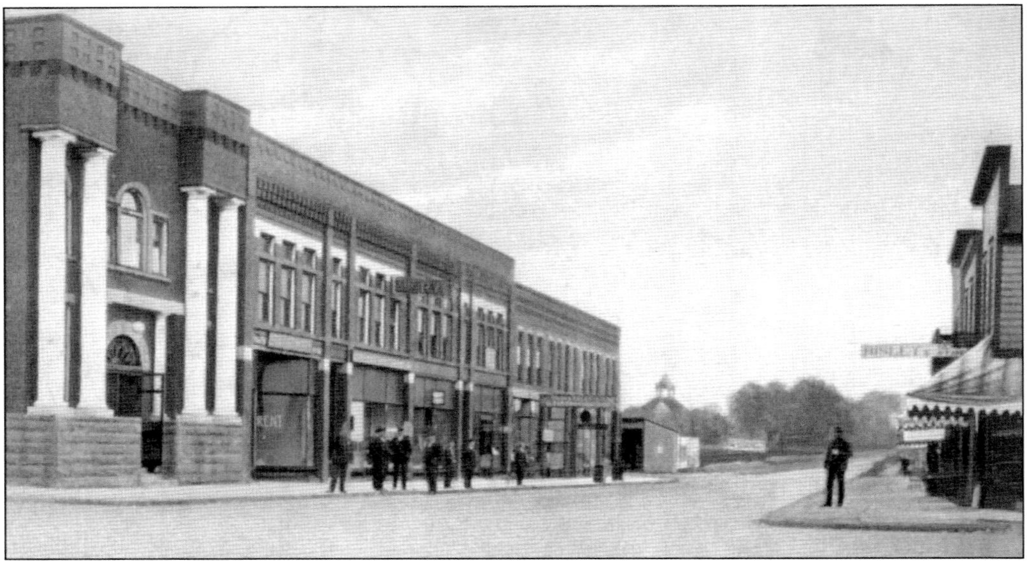

On the rebuilt southwest corner of Day Street and Park Avenue is the Conneaut Mutual Loan and Trust Company's harbor branch. The main office was located uptown in the Stanley block at Main and Washington Streets. On the left side of Park Avenue (above) are businesses, including a drug store, tailor, hardware store, and saloon. Also visible on Park Avenue are the Conneaut Electric Street Railway tracks. In the rebuilt brick building on Day Street (below) were the Jester Post Office, shoe store, clothing, and dry goods. Further down the street, there were six saloons, two hotels, two barbershops, and a grocery store.

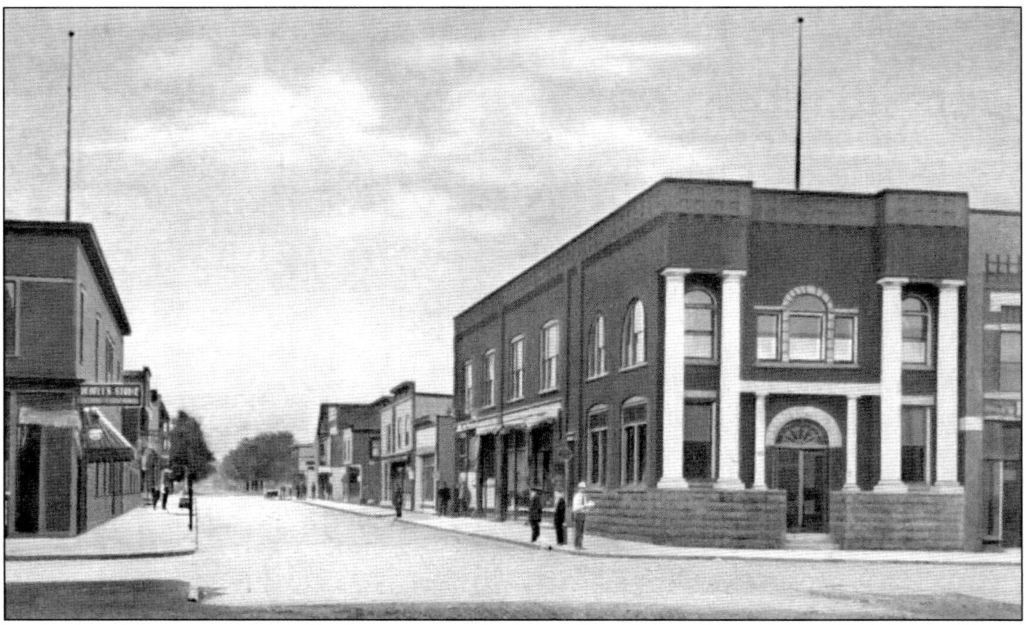

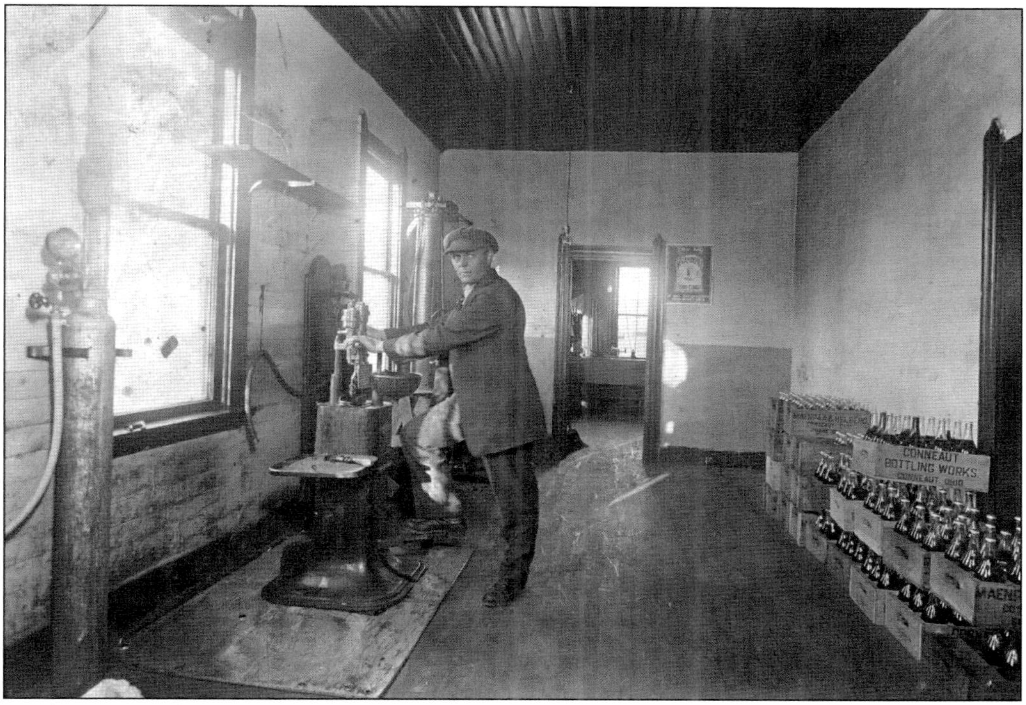

Jacob Maenpaa demonstrates a new machine at the Conneaut Bottling Works, located at 892 Day Street, Around 1925. Maenpaa was born in Finland on April 18, 1886. He came to the United States in 1909 and was employed by the Pittsburgh and Conneaut Dock Company. On January 14, 1913, he and Jacob Helberg began bottling a variety of soda pop flavors under the brand name Tuaennoc. (Jack Morton.)

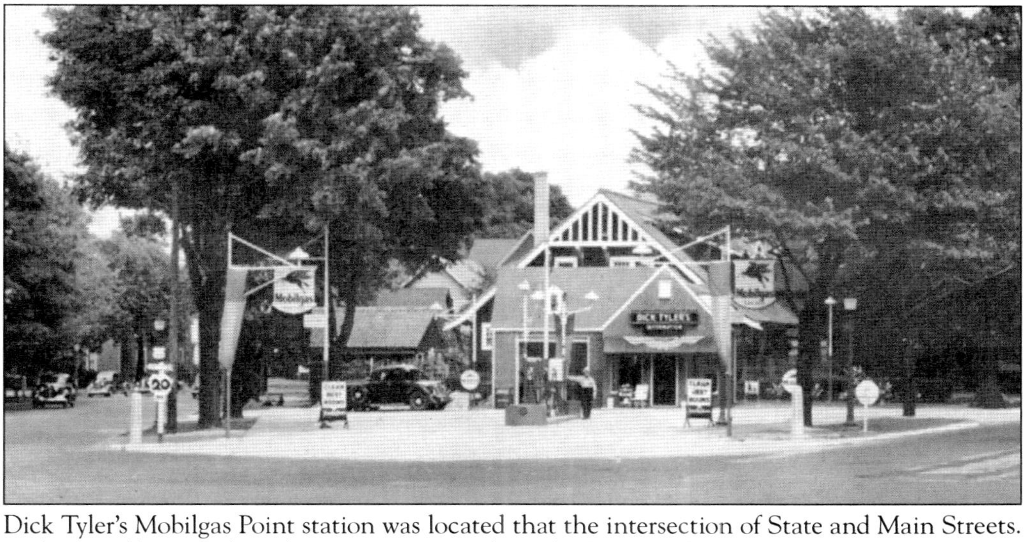

Dick Tyler's Mobilgas Point station was located that the intersection of State and Main Streets. Dick started his petroleum business in 1916 and was the first petroleum distributor in northeastern Ohio. He owned several gas stations in Conneaut and claimed many firsts, such as cleaning windshields and offering free air for tires. In his distribution business, he was the first to use a hose reel on the back of his truck. (Joyce Mackey Wright.)

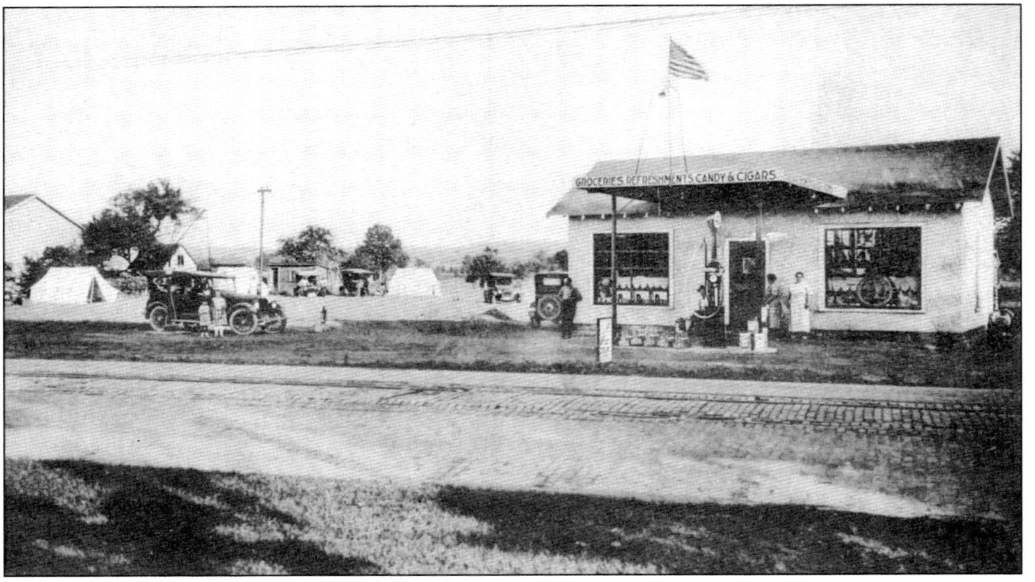

In the 1920s, Gaukel Brothers State Line Garage and Tourist Camp in East Conneaut was an oasis for travelers along U.S. Route 20. Services listed on the back of this postcard are free camping, restrooms, tents, cots, groceries, supplies, refreshments, showers, gasoline, oils, storage, accessories, free air, and sanitary conveniences. (Mary Jane Fritts.)

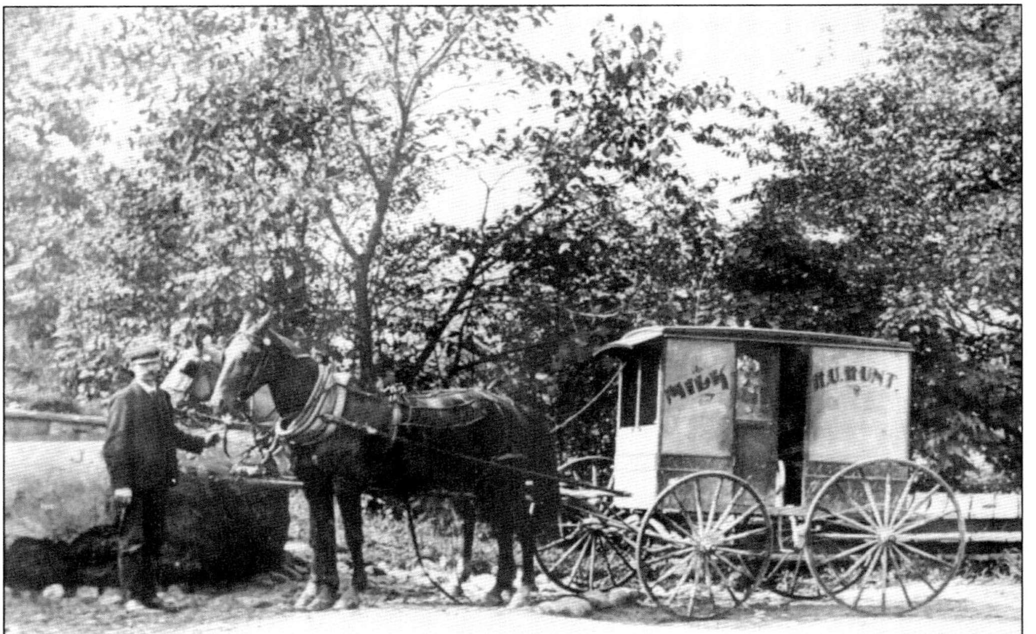

In 1896, J. B. Tyler hauled a 20-ton rock up 60 feet from the valley and set it on the west side of East Main Road hill in East Conneaut. After the stone was placed on a 10-cord stone foundation, a stonecutter spent 10 weeks shaping it into a concave watering trough for passersby. The rock was fed by a spring at the top of the hill.

For 14 years, television's Mr. Whipple chastised people for squeezing the Charmin. In the 1920s, Conneaut's real-life Mr. Whipple had a store on the southeast corner of Harbor Street and Nickel Plate Avenue. He was more concerned with people squeezing his tomatoes. Whipple's was one of eight small grocery stores on Harbor Street in 1920.

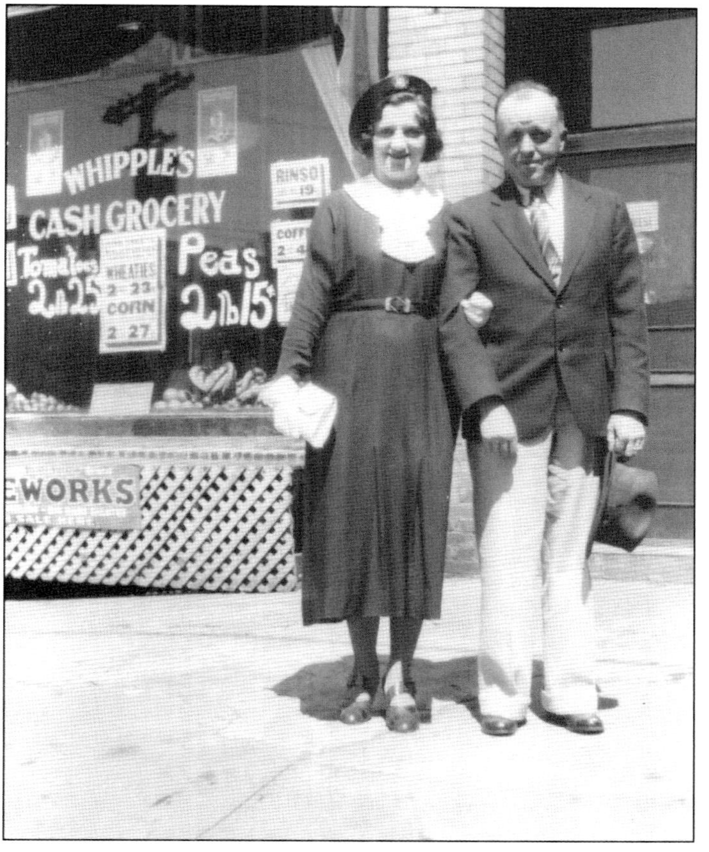

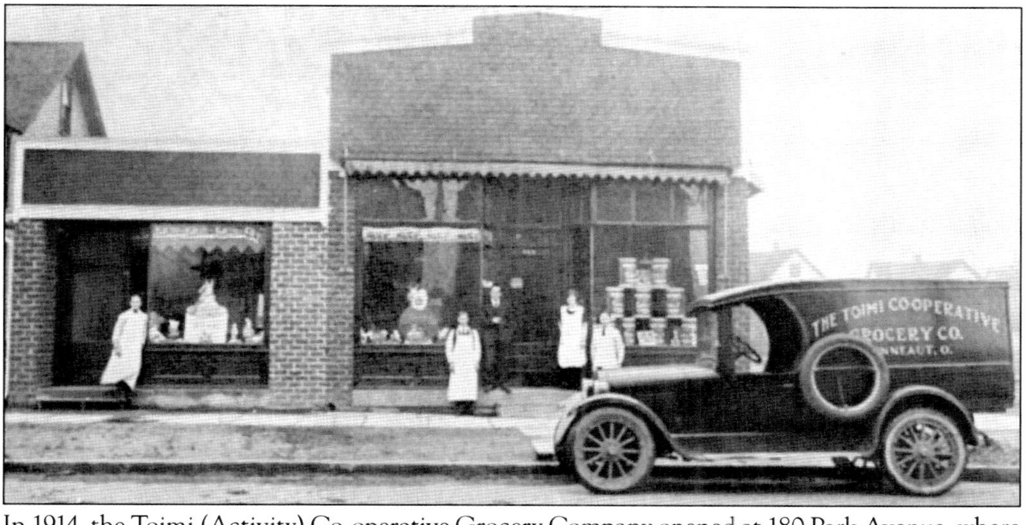

In 1914, the Toimi (Activity) Co-operative Grocery Company opened at 180 Park Avenue, where Finns could shop in their own store and speak in their own tongue. In the summer of 1919, the store moved into this building on the southwest corner of Broad Street and Erie Avenue. A meat department was added the following winter and a dairy in 1933. The business operated at this location until January 28, 1955.

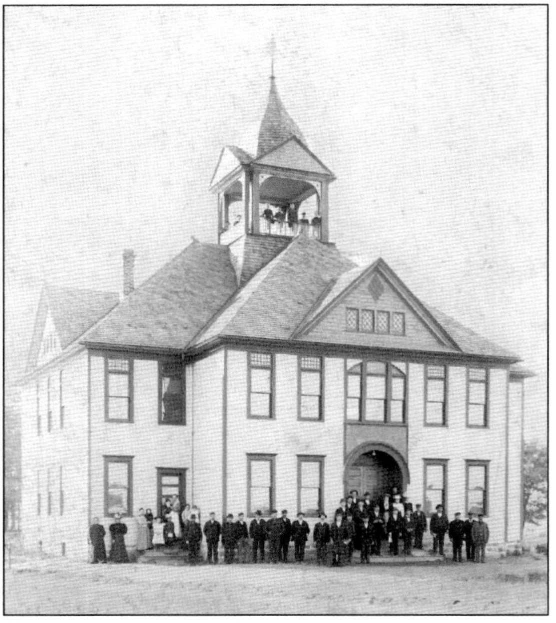

On the southwest corner of Buffalo Street and Park Avenue, Kilpi (Shield) Hall was built in 1899 by the Kilpi Temperance Society. The society existed from 1894 to 1976, with a peak membership of 140 in 1902. More than a crusading organization, it was also a Finnish fraternal, cultural, and social institution. Now the Conneaut Community Center for the Arts, the hall is listed on the National Register of Historic Places.

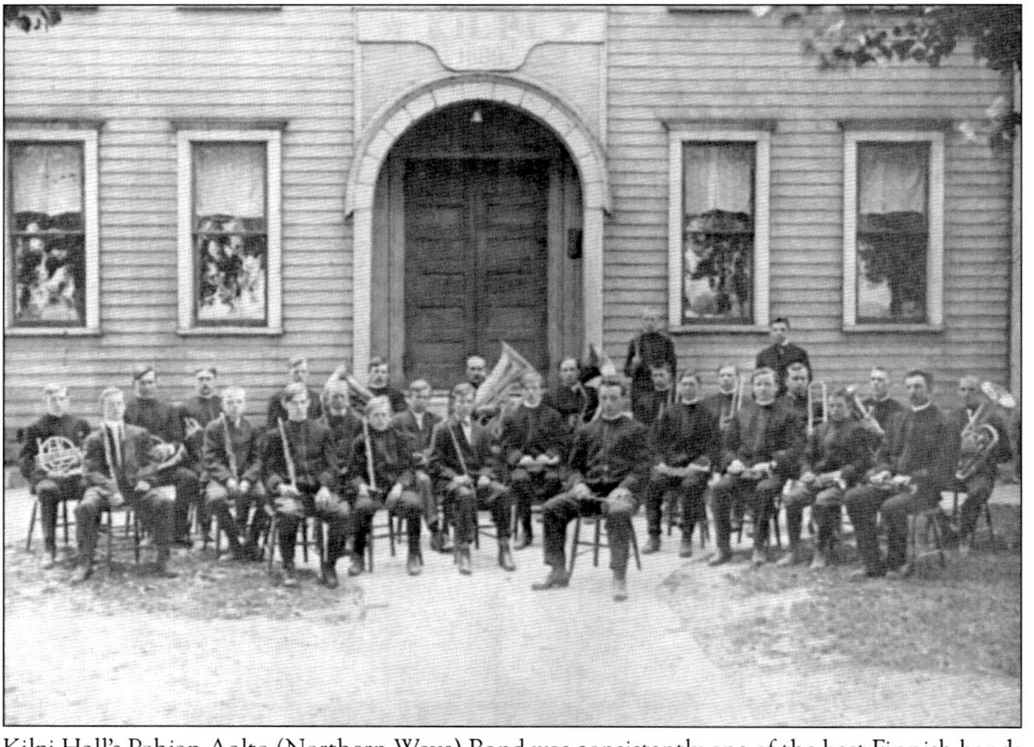

Kilpi Hall's Pohjan Aalto (Northern Wave) Band was consistently one of the best Finnish bands in the region. When the band was formed, the Kilpi Temperance Union provided the large instruments, and the Pittsburgh and Conneaut Dock Company provided the uniforms. In 1927, the Ashtabula Harbor Humina (Murmur) Band toured Finland under the baton of former Pohjan Aalto conductor George Wahlstrom.

Seven

EDUCATION

Salem's (Conneaut) first school was held in a log building with clapboard siding at the mouth of Conneaut Creek.

When Conneaut incorporated in 1834, the first school built was near the corner of Main and Washington Streets. Conneaut Academy incorporated on February 14, 1835, and held classes in a building at the corner of Main and Mill Streets.

In 1868, students from out of town would continue paying regular tuition. After leasing the building and grounds for 10 years, the academy was absorbed by the local school district.

In 1871, in addition to the departments in the high school, there was a grammar school (junior high) and four primary (elementary) schools.

A high school department was organized in 1878. The first class to graduate from the high school department was in 1880. The first commencement exercise was held, in the town hall assembly room, on June 17, 1880.

In May 1878, a school tax levy totaling $3,000 was passed, and $2,500 of that was allocated for teacher's salaries.

In 1886, commencement exercises were held at the town hall where the crowd was standing room only, with more than 200 turned away.

In 1890, there were 3,214 residents in Conneaut, which doubled the 1880 census of 1,256. In 1892, construction of Conneaut Harbor docks resulted in an influx of Finnish immigrants. By 1893, there were 150 Finnish families in Conneaut, leading to the construction of Harbor Street and Dean Avenue schools.

Born in 1891, Amy Kaukonen was the first Finnish student to finish high school. She graduated at the top of her class in Conneaut and the Women's Medical College in Philadelphia as well. In 1921, Dr. Kaukonen was elected mayor of Fairport, making her the first woman elected in Ohio and the second in the United States.

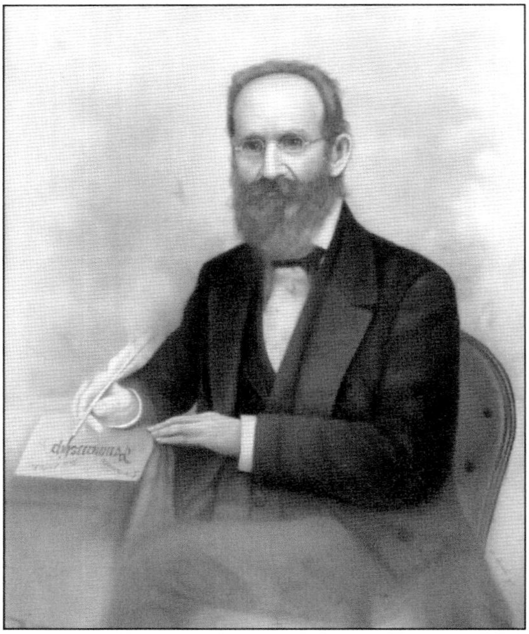

One of Conneaut's more successful early students was Platt Rogers Spencer, who was born on May 16, 1800. When he was nine years old, he moved to northeastern Ohio with his widowed mother, three brothers, and sister. He was formally educated at Harvey Nettleton's School in Salem. He went on to create the curvaceous Spencerian style of handwriting, which he said was inspired by Lake Erie waters, hills, and grapevines. (Geneva Public Library.)

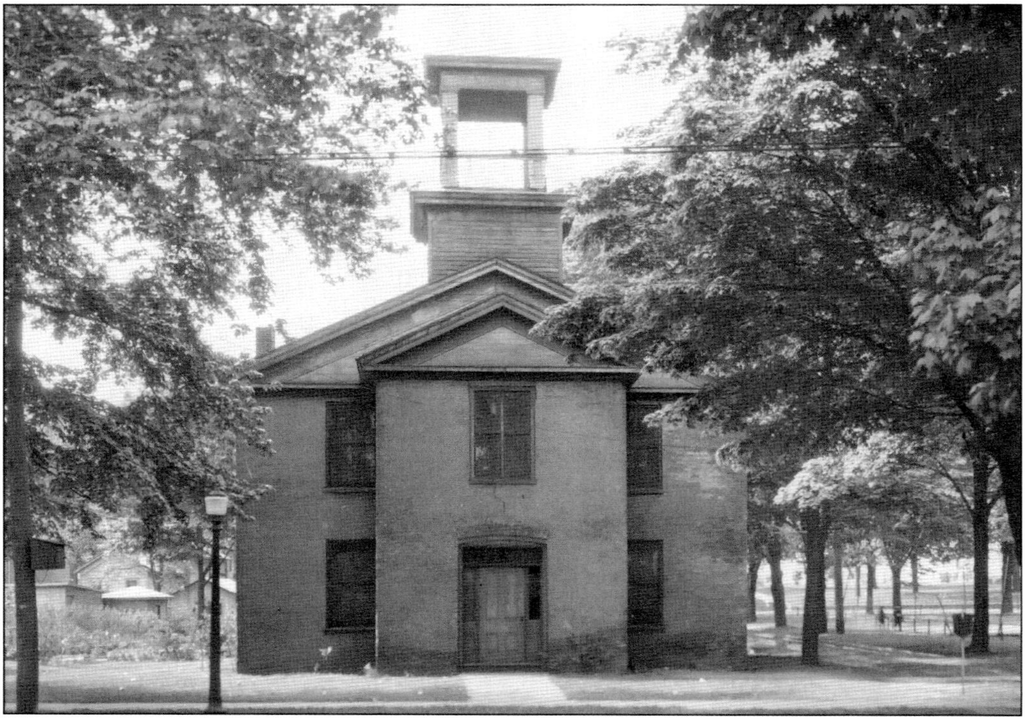

In 1845, the brick academy building was completed on the southeast of Main and School Streets. Students were charged tuition on a per-class basis. On August 17, 1868, the village board of education announced that it had made arrangements with the board of the academy to permit local students, when qualified, to attend the academy free of charge. The building was demolished in 1928. (Ashtabula County Historical Society.)

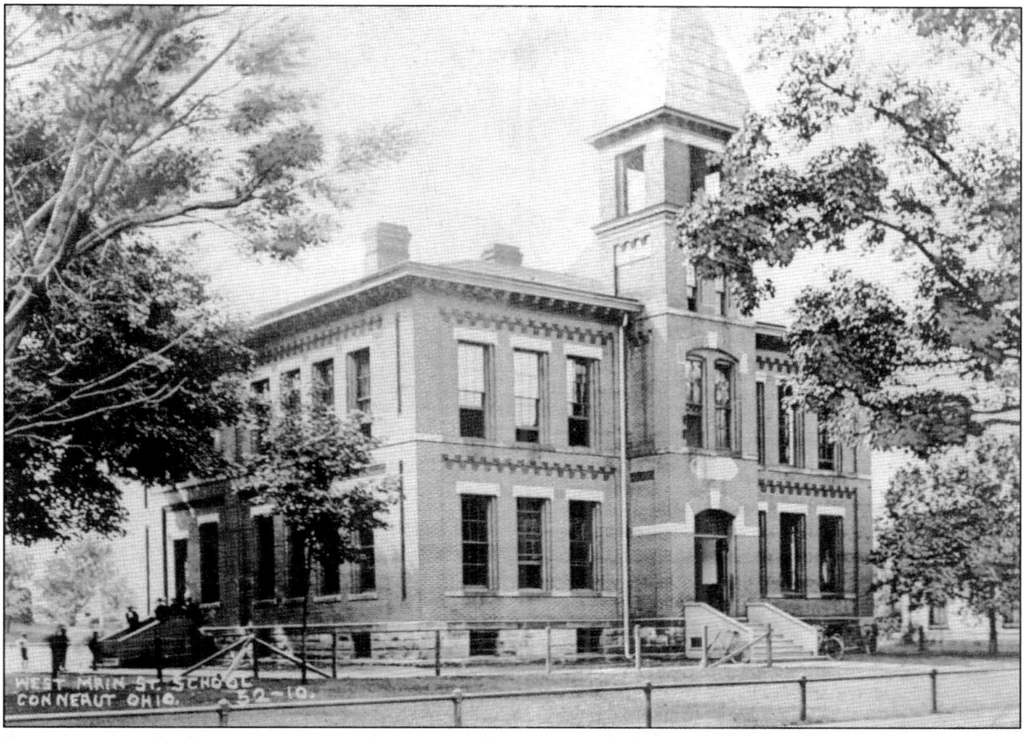

Beginning with the spring term of 1886, students attended the newly constructed Main Street School. The school, located on the southwestern corner of Main and School Streets, accommodated all high school and some grammar school students. By 1893, Main Street School had become overcrowded, forcing the high school to hold classes in the town hall. The Main Street School continued to be used as a grammar school until it was torn down in 1937. (Mary Jane Fritts.)

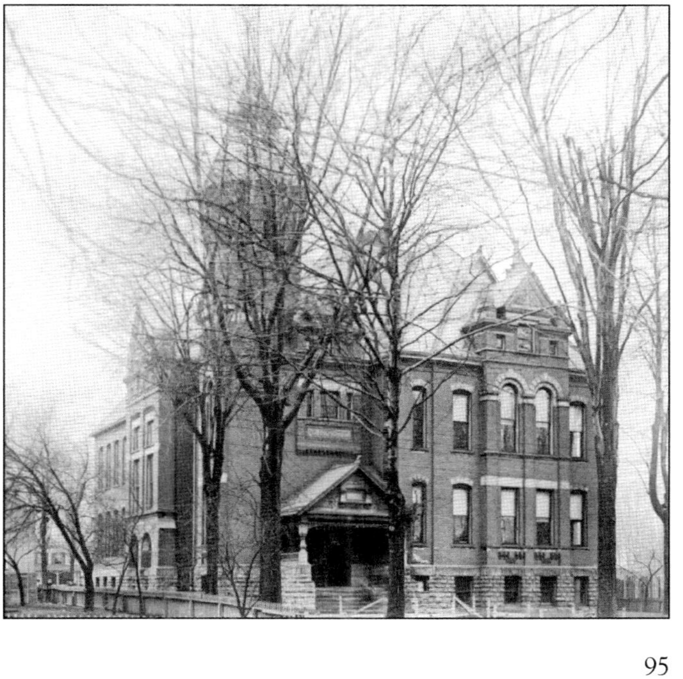

Broad Street School opened as a high school with several grammar school grades on September 11, 1894. Commencement exercises continued to be held at the town hall or the Weldon Opera House until 1910. The Weldon Opera House was located on the third floor of the Harwood block on the northeast corner of Broad and Main Streets. The school was demolished in 1960. (Mary Jane Fritts.)

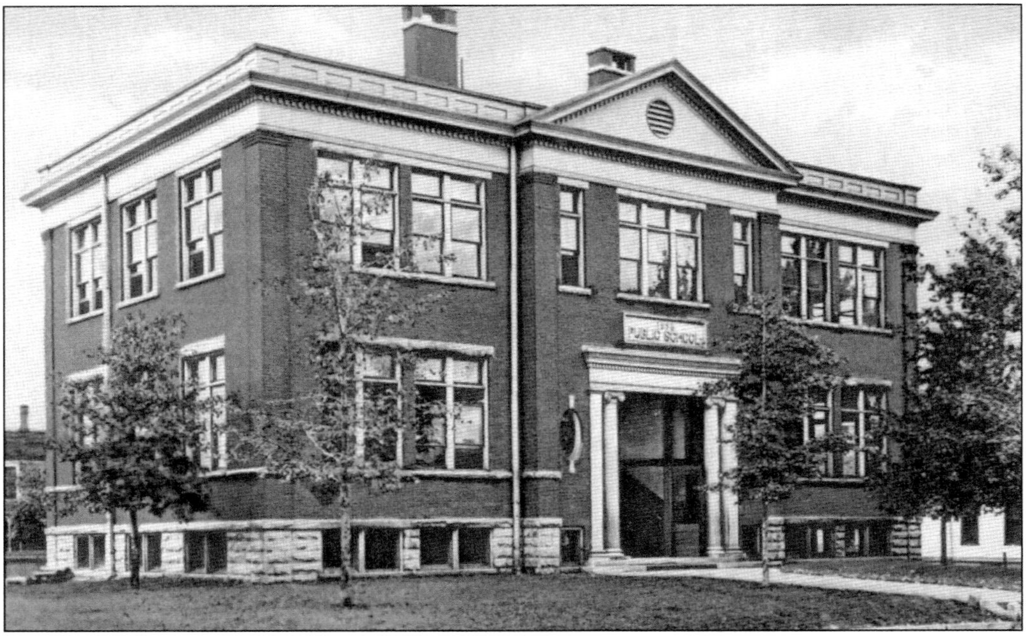

Fifteenth Street School was located on the northwest corner of Harbor and Fifteenth Streets. The school was known as Harbor Street School when it was built in 1898. In 1915, the Fifteenth Street entrance and another wing were added to the back of this building after demolishing the house behind the school (left). The school was demolished in 1962.

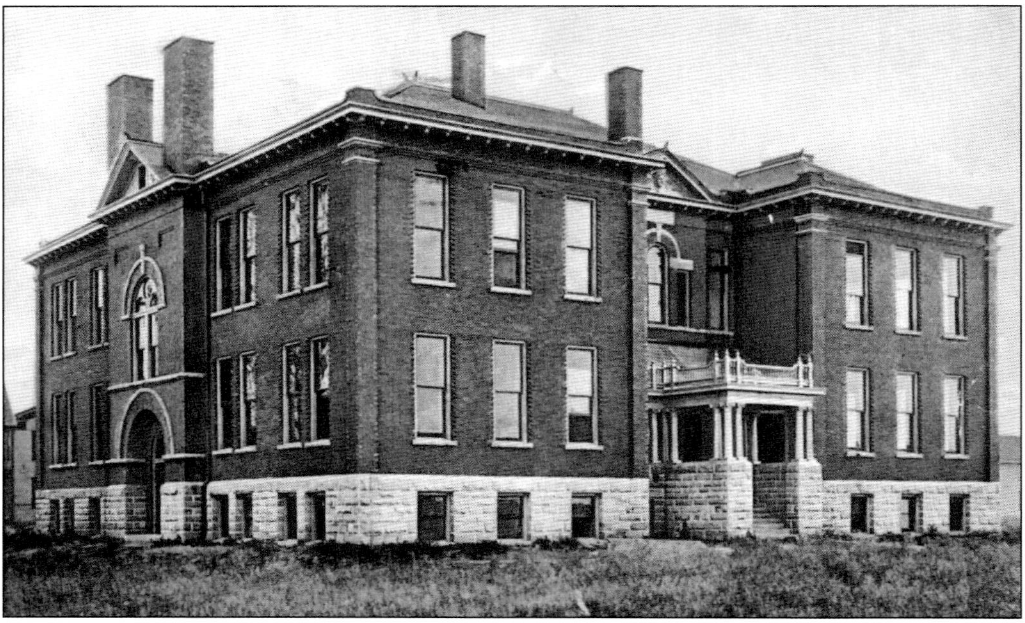

Dean Avenue School was located on Dean Avenue between Broad and Harbor Streets. The school was built in 1901 on the site of an older wooden schoolhouse, still visible to the left of the new brick building. In 1921, an addition onto the back of the building included a kindergarten on the first floor and a music room on the second floor. The school was torn down in 1969.

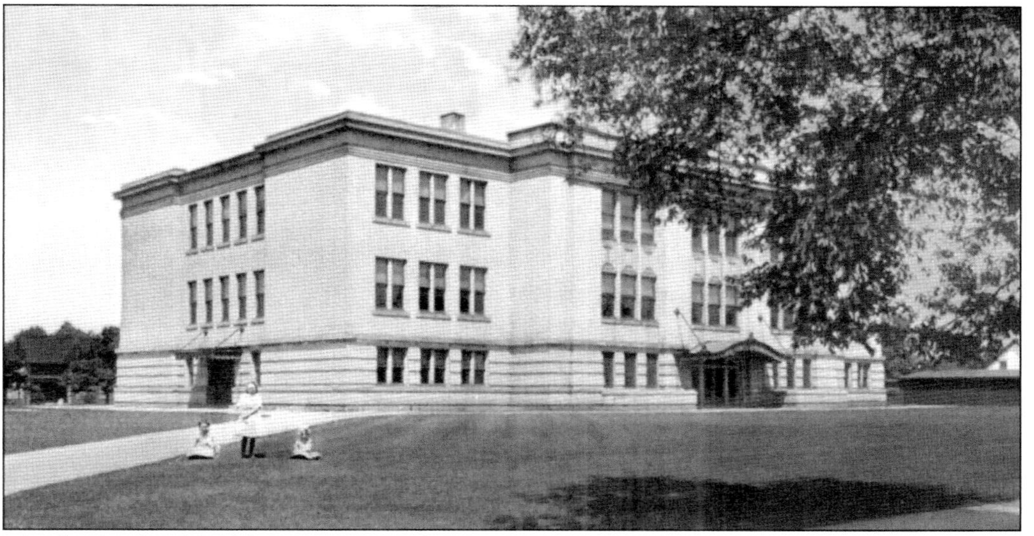

This high school building was situated on 2 acres of ground facing Sandusky Street between Adams and Jackson Streets. The building opened on September 3, 1909, and included a center stairwell with a skylight on the roof. Classrooms featured windows facing both the outside and the stairwell for additional natural lighting. It was demolished in December 1958.

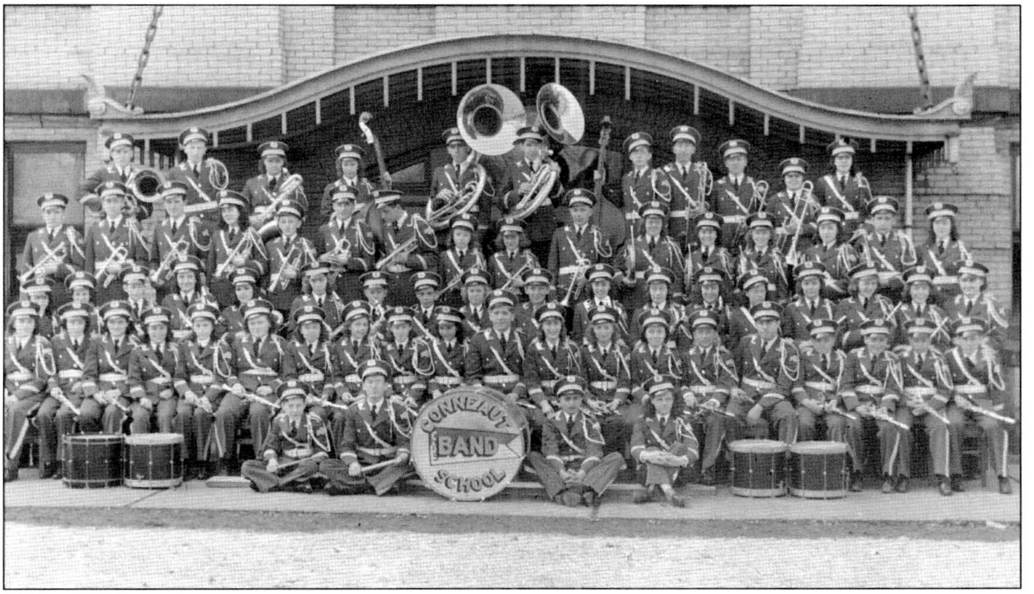

The 1941 Conneaut High School Band poses in their new military-style uniforms outside the Sandusky Street main entrance. The band was comprised of grades 7 through 12 under the direction of Kaarlo Mackey. Marching band practice began at 7:00 a.m. each weekday morning during football season. Concert band rehearsals were held after school.

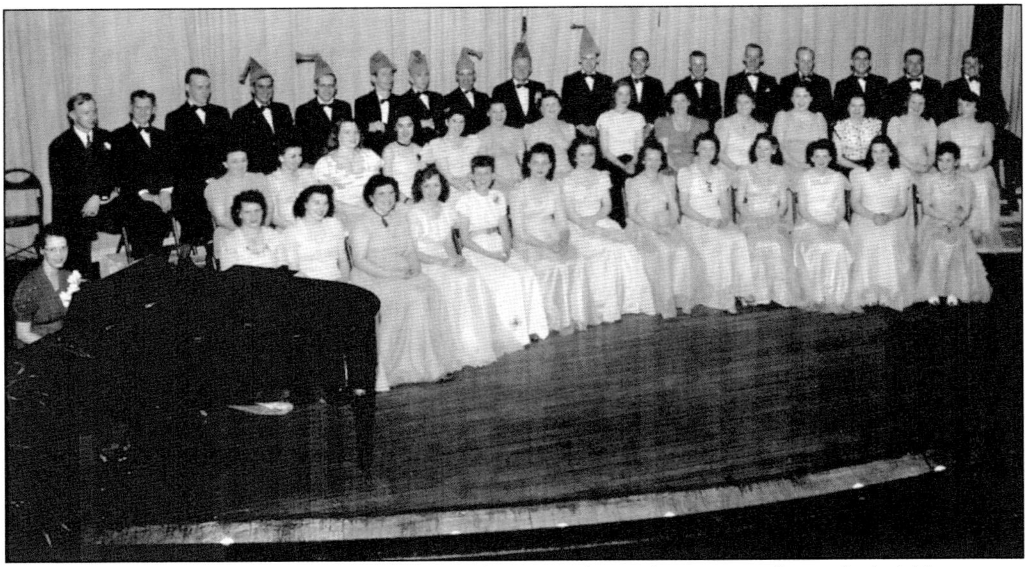

One of the most elegant features of this building was the auditorium with its plush folding seats and stage, complete with footlights and curtains. Not only were commencement exercises now held in the high school, but it was also used for plays and concerts by senior, junior, and elementary school bands. Community events, such as The Astatic Corporation Choir's Christmas concert and shows by traveling entertainers, were also held in the auditorium. (Karl A. Robotham.)

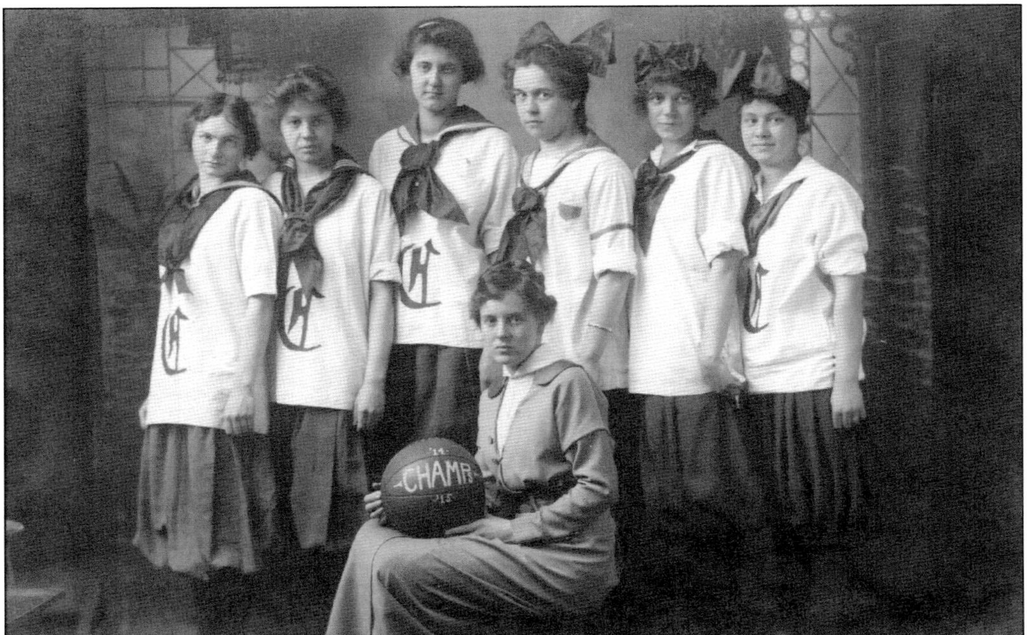

Athletic opportunities for girls were not provided in many schools until women's suffrage in 1920, the 1964 Civil Rights Act, or Title IX of the Education Amendments of 1972. Being more progressive, Conneaut High School boasted a girl's basketball team as early as 1902. In 1915, coached by Vera Webster, Conneaut won the Ashtabula County Championship. The one amenity, which this team lacked, was a girl's locker room and showers. (Mary Jane Fritts.)

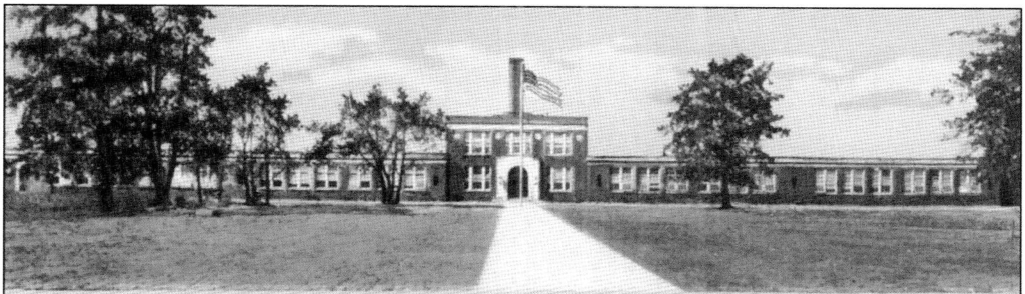

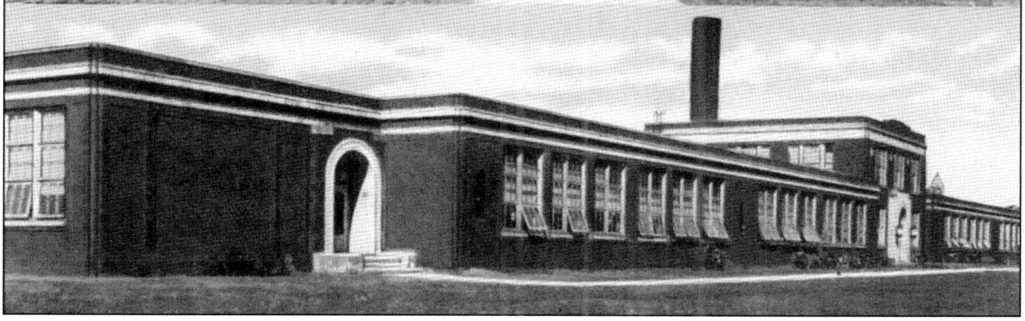

In 1923, the junior high school opened its doors to accommodate seventh, eighth, and ninth grades. The second floor band room was also used by senior high and elementary student rehearsals. In May 1952, an addition onto the back of the junior high included a machine shop and gymnasium, with locker rooms and showers for both boys and girls. High school students would walk across Mill Street to the junior high school for some classes.

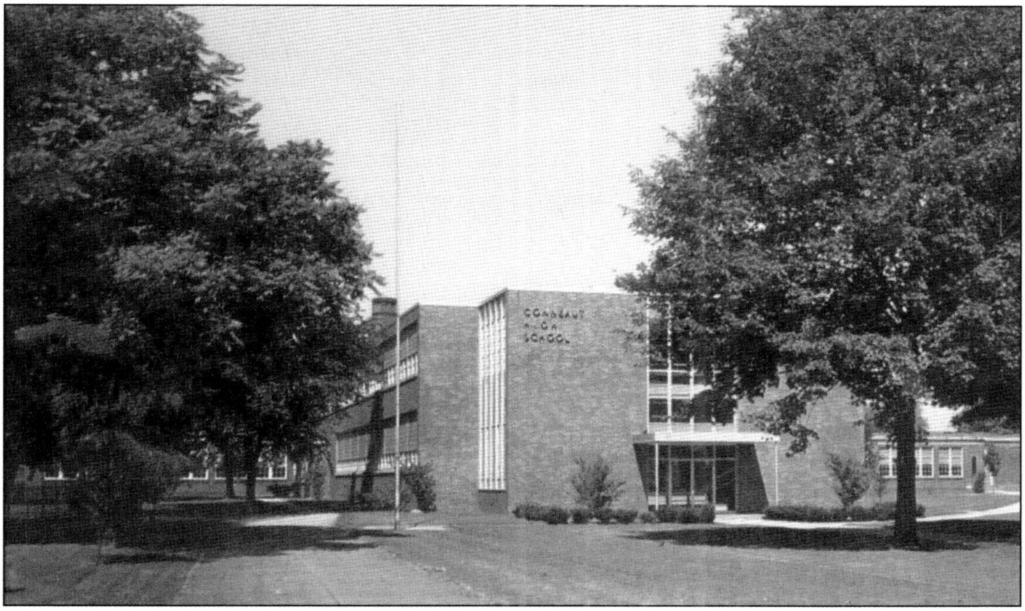

In 1958, this two-story addition was built onto the front of the junior high school, creating a new high school accommodating grades 7 through 12. In 1963, the City of Conneaut and Village of Lakeville merged. The following year, grades 10 through 12 attended high school in this building. Grades seven through nine attended junior high school in the former Rowe High School building on Rowe Street in East Conneaut.

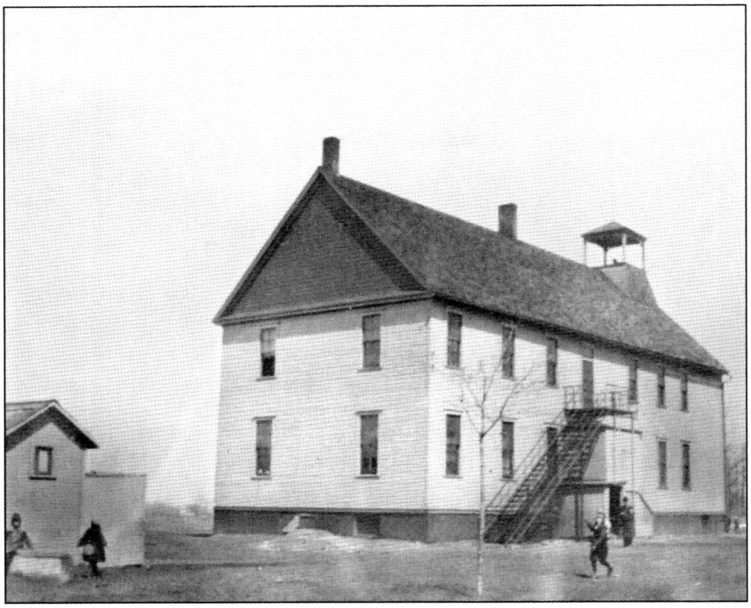

The Oklahoma School was located on the south side of Clinton Avenue between Chadman and Parker Streets. The building originally had only one floor. Students were not promoted to the Conneaut High School until 1903 when the entire eighth grade passed the Boxwell Examination, proving they were just as proficient as the rest of Conneaut's grammar school students.

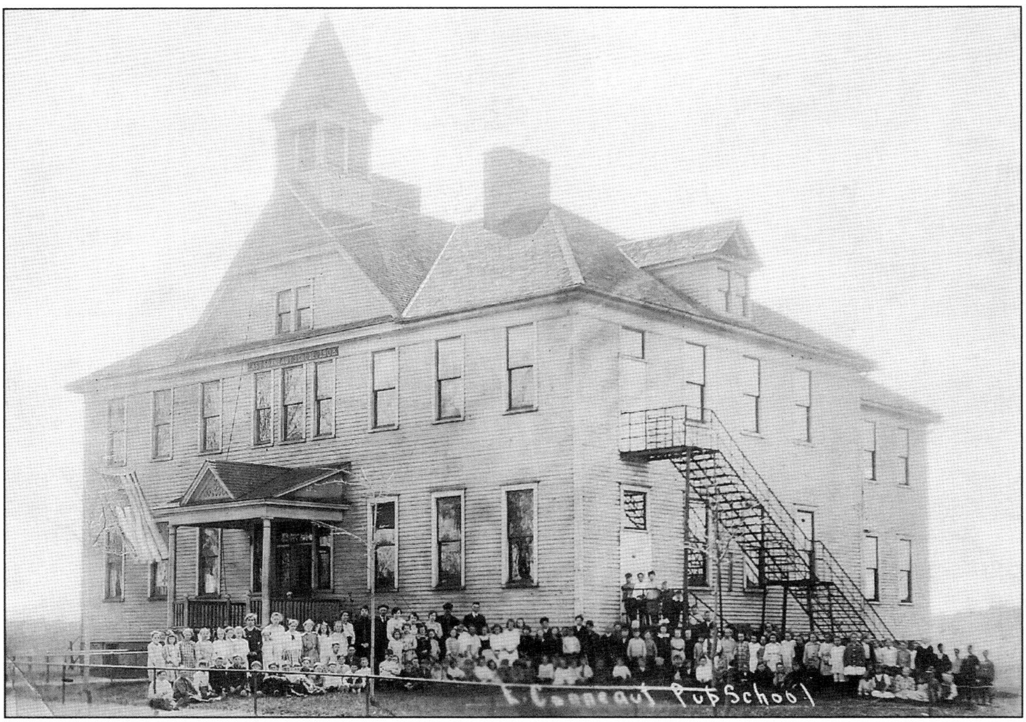

East Conneaut School, located on the southeast corner of East Main and Middle Roads, was originally a rectangular building with the portico on the front. By 1919, the tower, fire escapes, and a square addition onto the rear of the building were added. The building was demolished when the Conneaut Township High School on Rowe Street was enlarged to accommodate the lower grades.

Eight

Religion

One of Conneaut's early residents stirred up a religious hornet's nest, which continues to be hotly debated almost 200 years later.

In 1761, Solomon Spaulding was born in Ashford, Connecticut. He served in the Revolutionary Army, studied law, and graduated from Dartmouth in 1785. Following graduation, he became a minister in New England.

In 1800, Aaron Wright was credited with discovering the "Conneaut Giants." Wright found small burial mounds, which contained the remains of a large-boned race of people, he believed pre-dated the Eriez and Iroquois Indians. The skeletons were blackened and disintegrated when unearthed and exposed to air. The skulls and jaws were said to be large enough to cover a man's head and face.

In 1809, Solomon Spaulding moved to Conneaut and established an iron foundry, with Henry Lake, which failed. In order to make ends meet, Spaulding decided to write a fictitious novel based on the Conneaut Giants. He read his work, *Manuscript Found*, to friends but was unsuccessful in having it published by Patterson and Lambdin in Pittsburgh.

Before he died in 1816, Spaulding told his wife and friends that a printer, named Rigdon, had copied his manuscript while it was in Patterson's office. Shortly after Spaulding's death, Sidney Rigdon became a preacher and wandered about Pennsylvania preaching strange unorthodox doctrines. In 1826, he became a Campellite preacher in Mentor, Ohio.

Joseph Smith said he was guided to golden plates by the angel Moroni on September 22, 1823, near Palmyra, New York. From 1827 to 1829, he translated the golden plates into the *Book of Mormon*.

Upon hearing a Mormon elder's lecture on the *Book of Mormon*, Conneaut residents familiar with Solomon Spaulding's work suspected that it was based on *Manuscript Found*. John Spalding and others later testified to the similarity between the two works.

The Mormon position, described in *The History of the Reorganized Church of Jesus Christ of Latter Day Saints*, was that Aaron Wright, Henry Lake, and others were "endeavoring to invent a theory that would account for the origin of the *Book of Mormon*, other than the truthful narrative given of it, and in a way that would tickle the fancy and please the ear of the fable-loving."

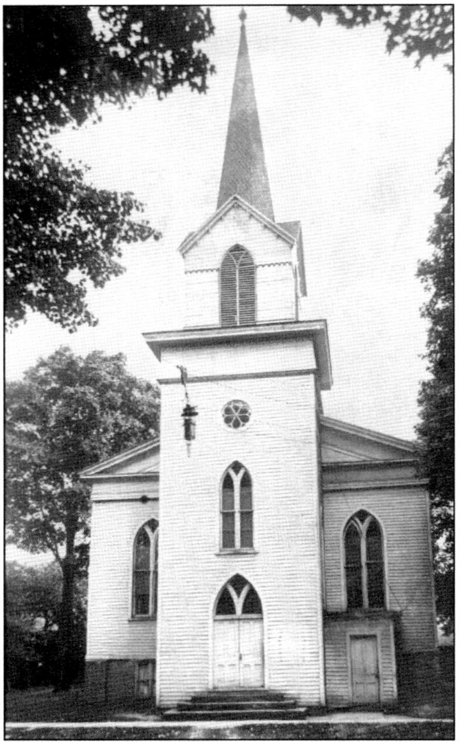

In 1834, the First Christian Church was built on the north side of Main Street between Cummins and Grandview Avenues, a location known as "The Center." Seven years later, James Pullman of Dunkirk, New York, was contracted to move the church 1 mile to 236 Buffalo Street using a rail system and horses. In 1913, the original church was replaced by the current brick structure that became the Free Methodist Church.

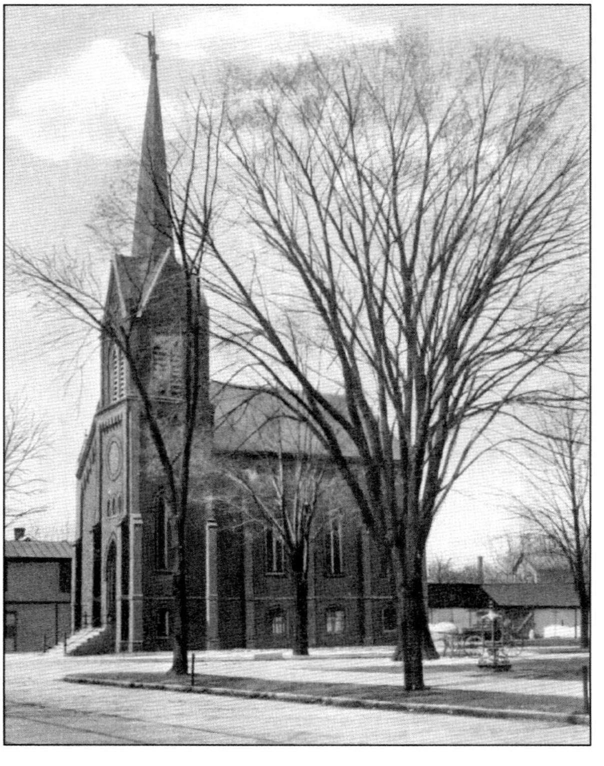

The Congregational Church of Conneaut was organized on April 14, 1849. The services were initially conducted by itinerate preacher Joseph Badger. The first church building, known as the "Old Brick Church" on Liberty Street, was completed in 1828. The building to the left, situated on the northwest corner of Main and Buffalo Streets, was finished in 1876 at a total cost of almost $20,000.

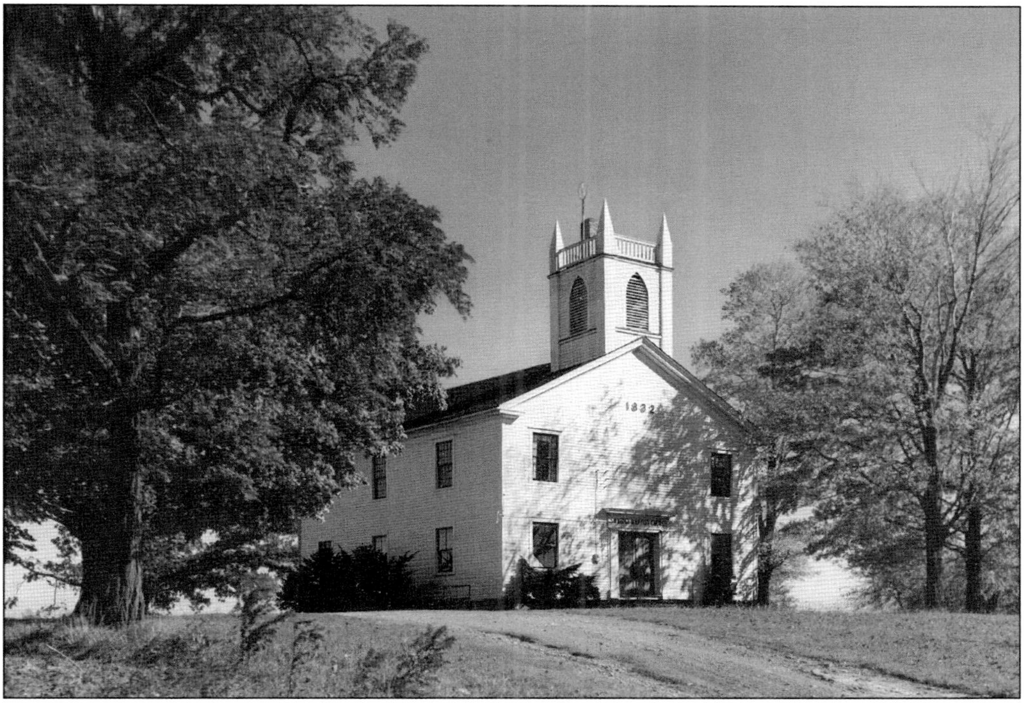

The Free-Will Baptist Church was organized on December 30, 1826. In 1832, the church building was erected, at South Ridge, for $2,000. The longest pastorate was that of Rev. Rufus Clark, who served 13 years. Reverend Clark and members Hiram Lake and Marshall Wright were part of the South Ridge Anti-Slavery Society. All three boldly used their homes as stops on the Underground Railroad during the Civil War.

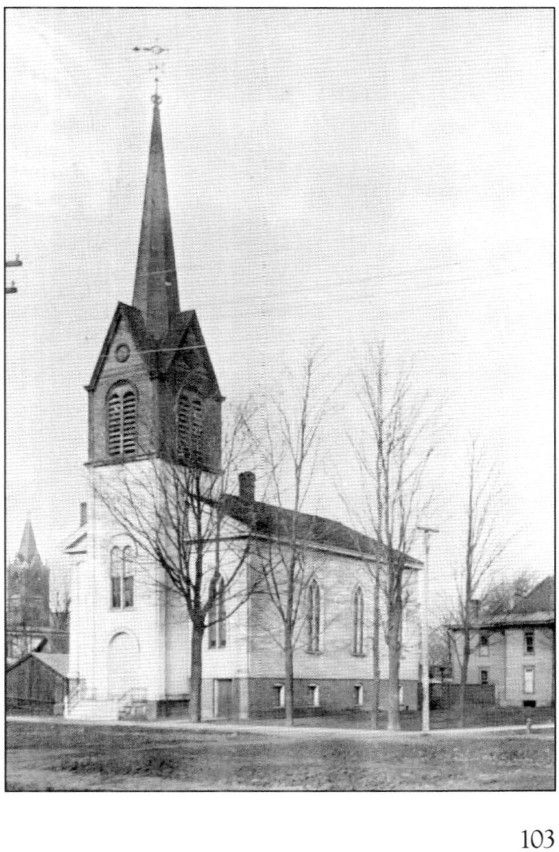

Conneaut's First Baptist Church was organized on October 18, 1831, in the old schoolhouse on the south ridge with 23 members. The church continued to meet at the schoolhouse on south ridge until the spring of 1837, when it moved to Conneaut village. In 1842, the congregation began building the church on the left on the northwest corner of State and Broad Streets. The church was destroyed by fire in 1900.

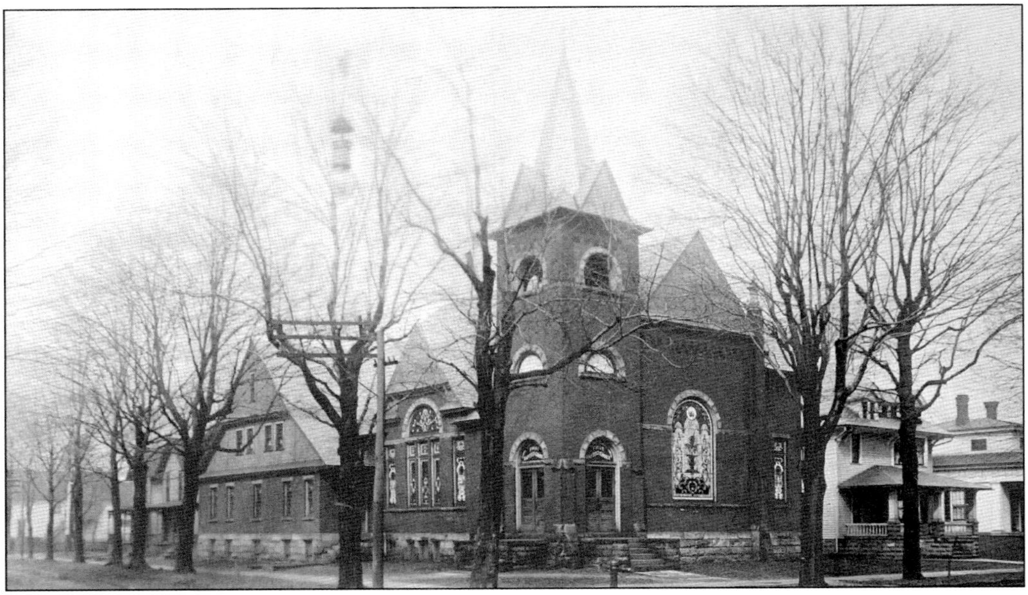

Following the 1900 loss of their church building, the First Baptist Church congregation constructed a new church on the northeast corner of State and Mill Streets. The parsonage is located to the right of the church. The future First Church of Christ Scientist is located to the right of the parsonage. (Mary Jane Fritts.)

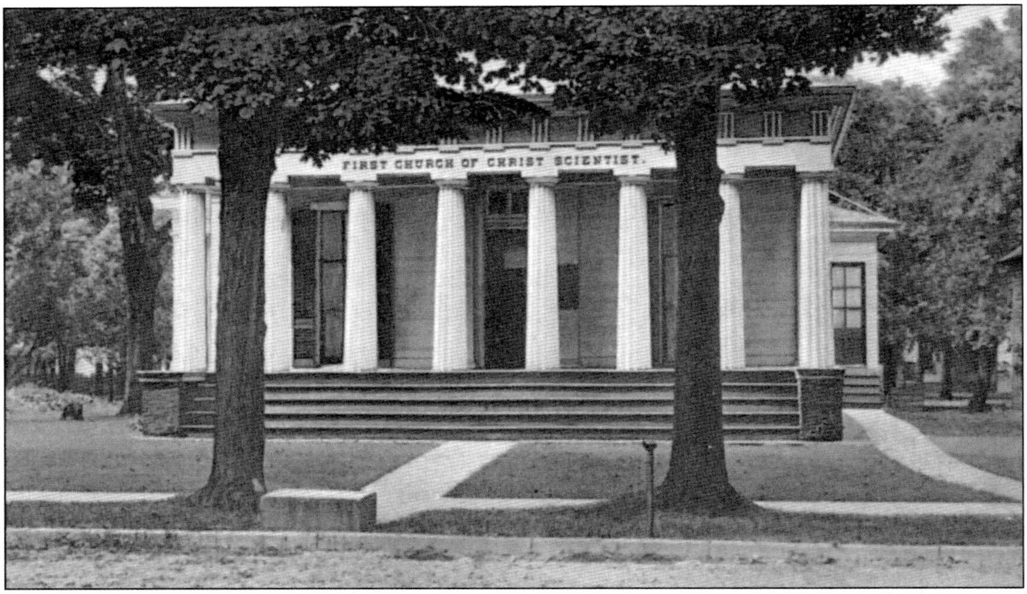

In 1879, the Church of Christ Scientist was founded in Boston. By 1905, the First Church of Christ Scientist had been established in Conneaut. By 1909, the church had acquired the former home of Calvin Wallingford at 352 State Street (pictured). The building is now the Baptist Youth Center. (Mary Jane Fritts.)

May Ellidia Record was born in Conneaut on November 2, 1868, to George J. and Mary J. Record. She was known to be intelligent, quick witted, and well educated. On November 5, 1890, May married Walter T. Findley, son of the pastor of the Congregational church in Conneaut. In 1893, she became ill with tuberculosis and died at Grace Hospital on January 18, 1905, at the age of 36.

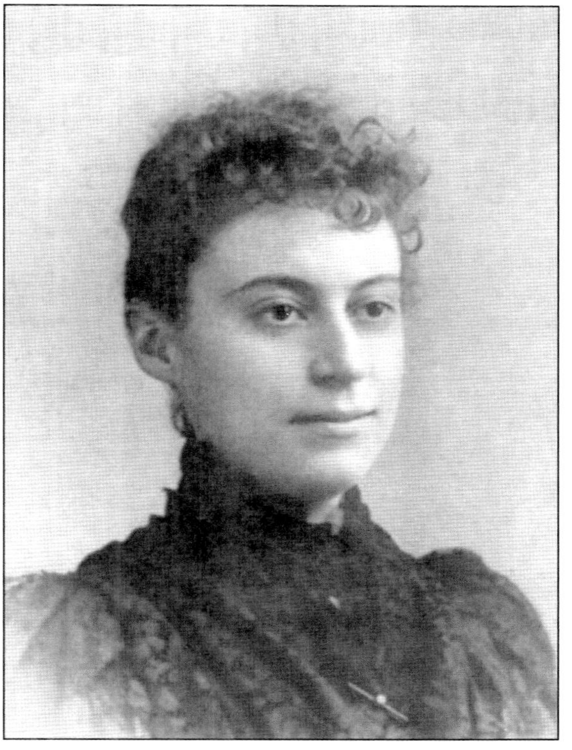

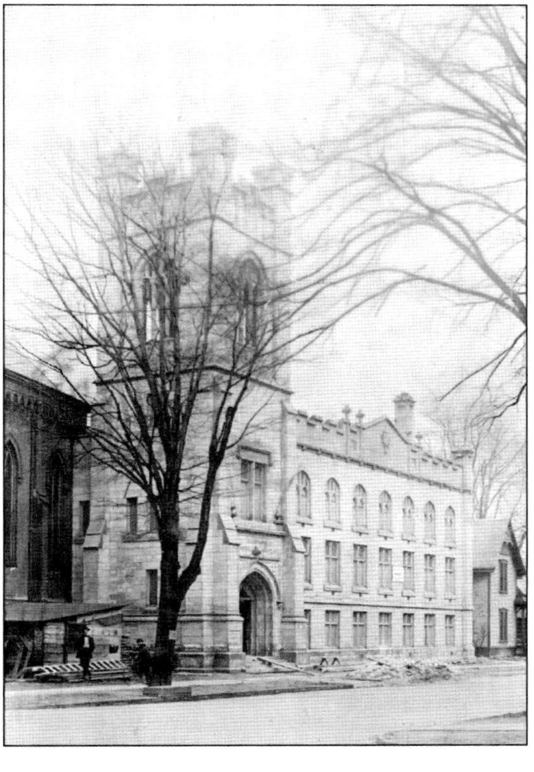

As recognition of the exceptional life of their only child, George and Mary Record planned and financed 100 percent of the construction of the May Record Findley Memorial. The cornerstone was laid on June 9, 1907, and the chapel was dedicated on June 20, 1909. With its bronze front doors, New Bedford Oolitic stone exterior, and interior lined with choice Vermont marble, it is the most elegant building ever constructed in Conneaut. (Mary Jane Fritts.)

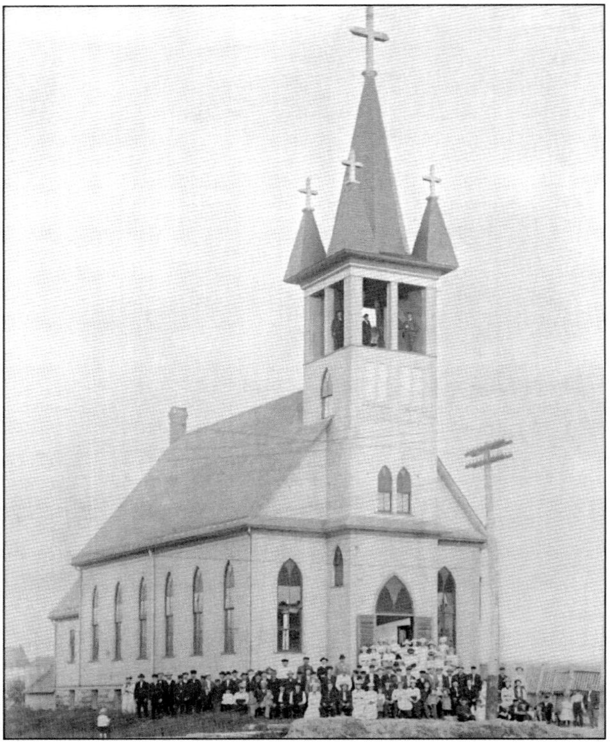

On July 14, 1895, the Conneaut Finnish Lutheran Evangelical Congregation was formed with 58 initial members. The Finnish Lutheran Church sanctuary, on the southeast corner of Broad and Erie Streets, was dedicated on July 21, 1901, with the church being built by mostly volunteer labor. The design was copied from a magnificent church in Ylistaro, Finland, on the southwestern coast, the original homeland for most of Conneaut's Finnish immigrant population. (Joyce Mackey Wright.)

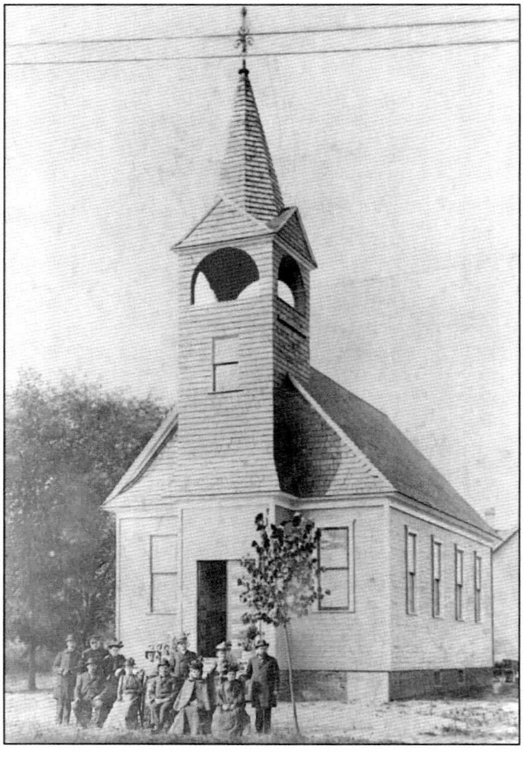

The Finnish Congregational Church was organized with nine founding members who built this church on the northwest corner of Broad Street and Lake Road at a cost of $1,793.78. The church membership never exceeded 28, which led to a hand-to-mouth existence throughout its brief history. The church was dedicated on September 30, 1900, and closed in 1911.

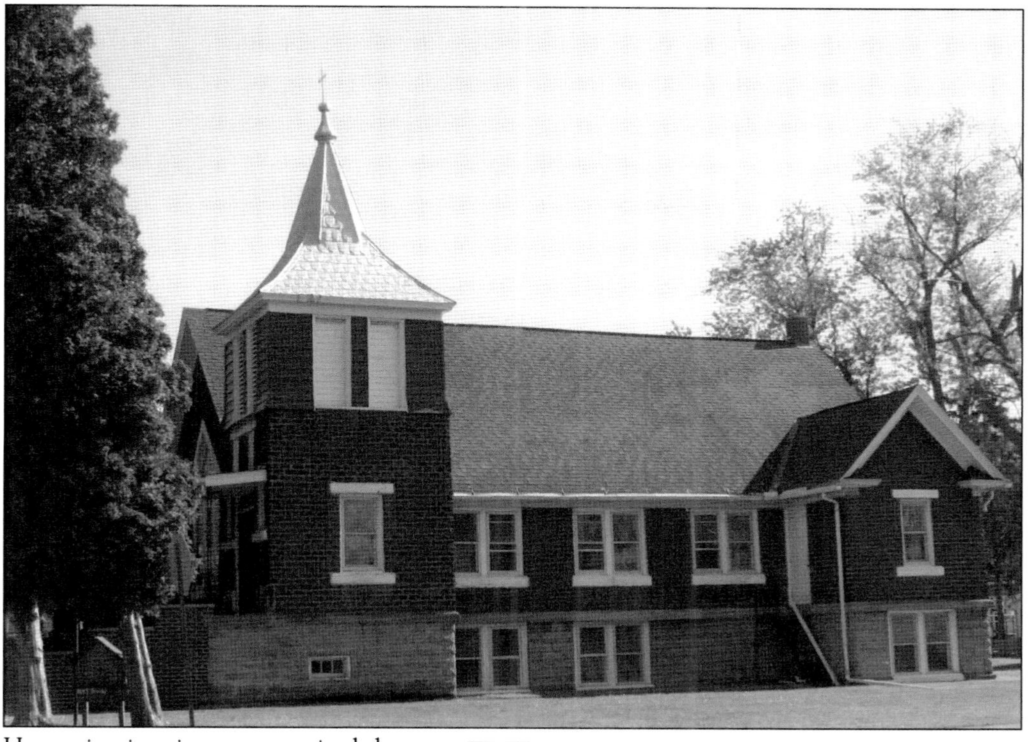

Hungarian immigrants organized the First Magyar Reformed Church in 1904. They held services in various churches throughout the city until they were able to construct their own building. In June 1912, the congregation acquired the property on the southwest corner Harbor Street and Lake Road for $400. In 1914, the cornerstone was laid and the church was completed the same year. (Steve "Pechew" Horvath.)

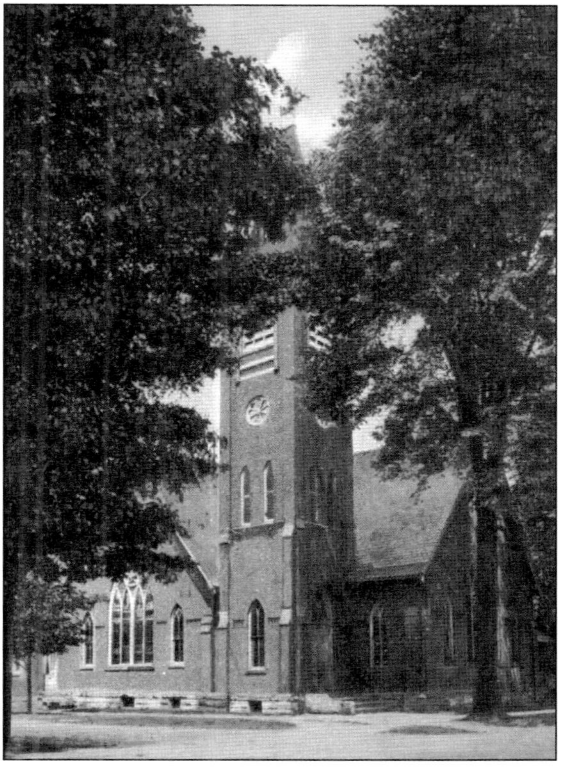

The Methodist Episcopal Church (later First Methodist Church), located on the northwest corner of Buffalo and Madison Streets, was built in 1887. A Methodist class was formed in Conneaut village about 1827 or 1828. One had previously been organized in the eastern part of the township and one at Amboy in the 1823.

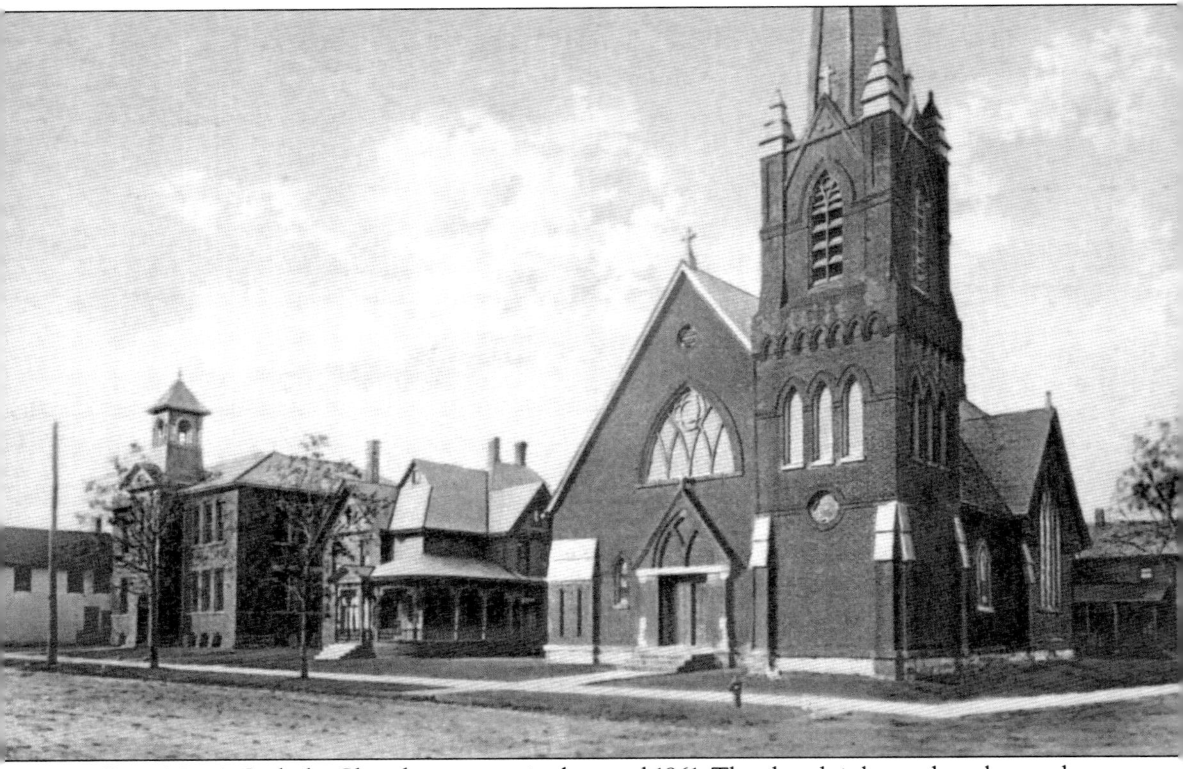

Saint Mary's Catholic Church was organized around 1861. The church is located on the northwest corner of State and Chestnut Streets. The building immediately to the left is the rectory, and the structure further to the left is the school that housed grades one through eight. The building behind the church is the nun's residence. (Mary Jane Fritts.)

Nine

Surrounding Area

When the Moses Cleaveland party surveyed the Western Reserve, the territory was laid out in 5-mile-square townships, with the exception of the ones bordering on Lake Erie. The original northern boundary of Monroe Township was Ohio State Route 84.

In 1799, Col. Stephen Moulten followed the Girdled Road trail from Pennsylvania along Underridge Road to today's Kelloggsville, becoming its first settler. The community was known as Ferguson's Settlement until 1814, when Amos Kellogg became Monroe Township's first postmaster. When Kellogg set up the post office in his house, the community became known as Kelloggsville.

On July 4, 1818, representatives of the villages organized Monroe Township and annexed the 2 miles north of Route 84, extending the boundary to the present-day Conneaut City southern boundary. The area was named Monroe Township in honor of then Pres. James A. Monroe.

By the 1830s, Kelloggsville had become a thriving community. Caleb Blodgett built many of the businesses around the village, including the Old Brick Tavern. Constructed in 1824 as Monroe Township's first brick building, it still stands today as a well-maintained home on the southeast corner of Monroe Center and Stanhope-Kelloggsville Roads. In 1834, when Blodgett sold his business interests and moved to Wisconsin, businesses and homes lined both sides of today's Stanhope-Kelloggsville Road from Bushnell Road to Gageville Road.

In 1805, Capt. Walter Fobes journeyed from Norwich, Massachusetts, and became the first landowner to settle in a community, which became known as Fobes Dale. Later settlers scoffed at the name, so it was changed to Norwich. This was also cause for disagreement, so the consensus of the townspeople was that a new name must be found. An itinerant named King, having no interest in the community, said he would give the townspeople four gallons of whiskey if they would name the town after him.

The village of Kingsville was organized in 1810. The business center was known as Kingsville Village, with Kingsville Township running from the village north to Lake Erie. In 1910, Kingsville Township incorporated as North Kingsville, with its southern boundary becoming the Nickel Plate Railroad.

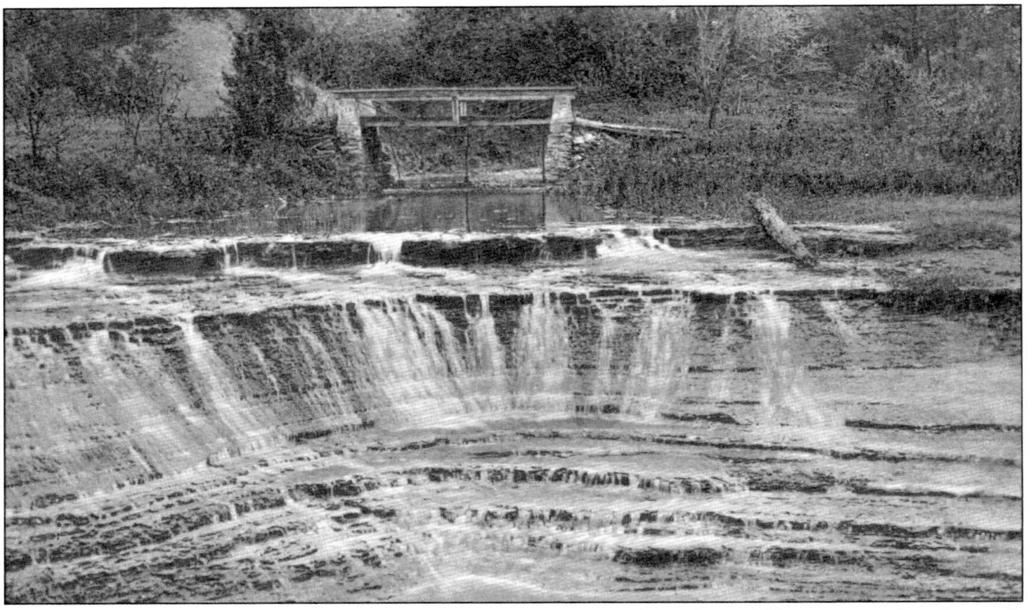

In 1807, Jacob Paden built a gristmill and saw mill, powered by these falls on State Road, south of Hatches Corners Road. Contrary to the postcard's caption, the falls are approximately 1 mile southwest of Tinker's Hollow. In this scenic corner of Monroe Township, a covered bridge lies less than a quarter mile north of the falls on State Road. (Mary Jane Fritts.)

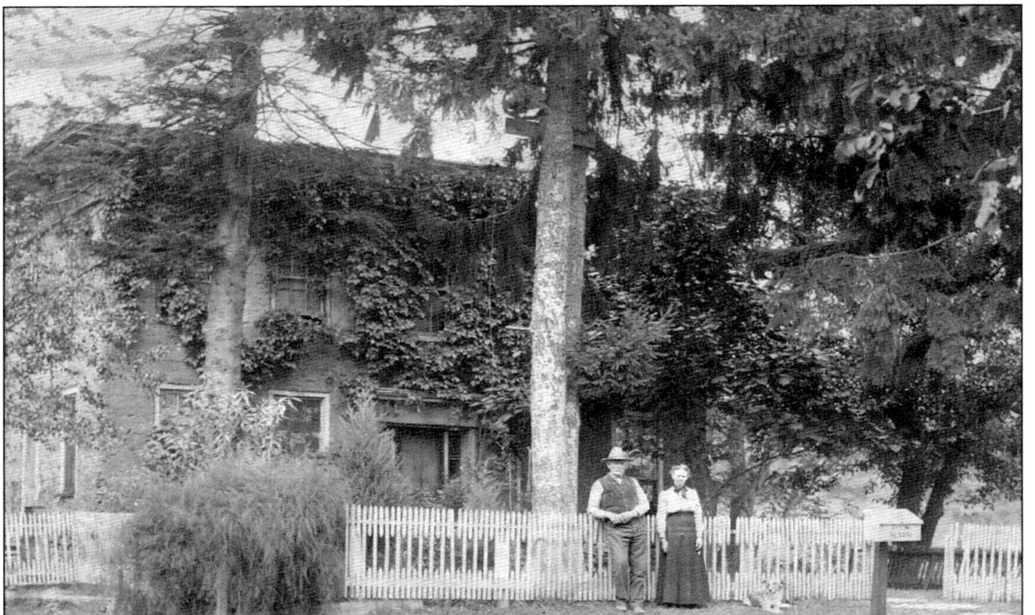

Julius and Susan Tinker pose with their dog Ginger outside the Tinker's Hollow family home in this *c.* 1920 family photograph. Julius was five years old when his father first demonstrated his mowing machine. Susan was known to have also worked in the family's foundry. Their son William Julius, born in 1872, was the last of Tinkers to work in his grandfather's foundry.

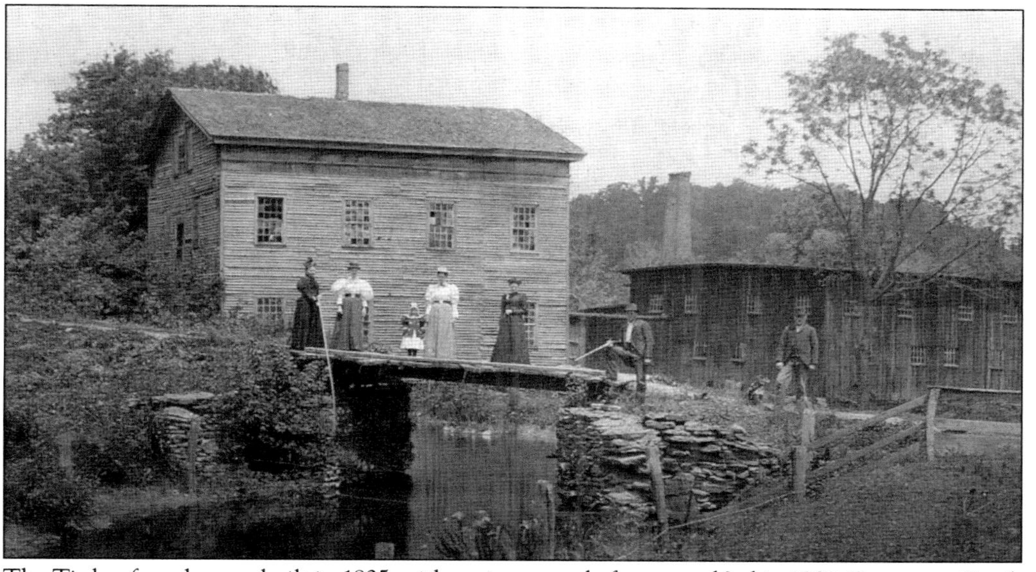

The Tinker foundry was built in 1835, with an iron cupola furnace added in 1839. Conneaut Creek was diverted to provide a millrace to power the machinery. Up to 22 men were employed in the manufacturing of mowing machines, sinks, skillets, kettles, stoves, and other tools.

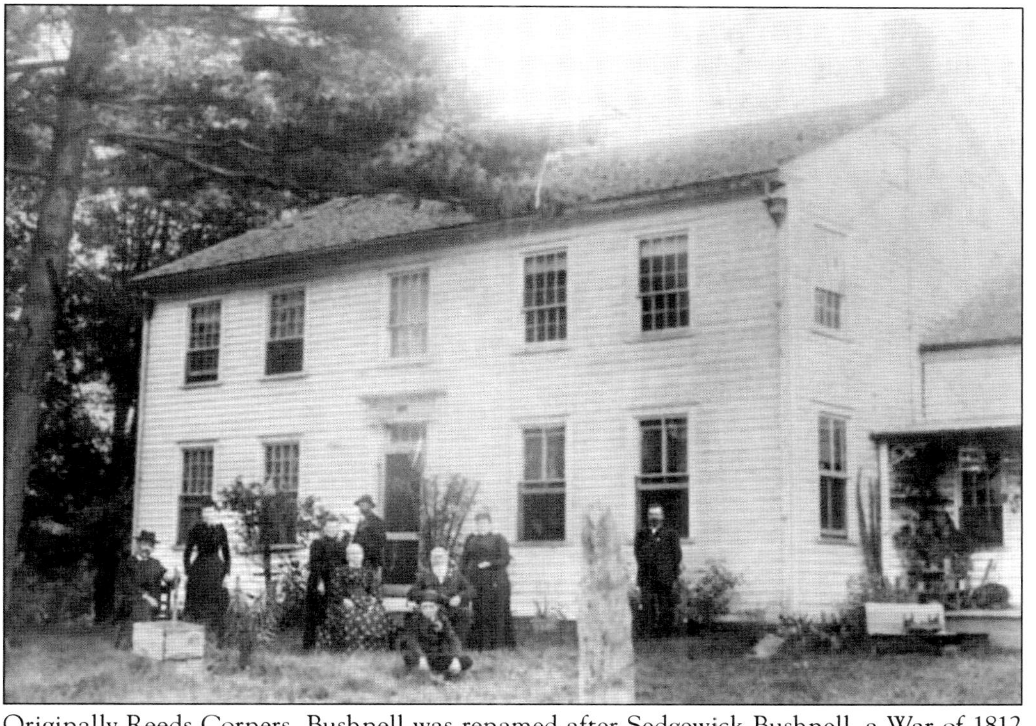

Originally Reeds Corners, Bushnell was renamed after Sedgewick Bushnell, a War of 1812 veteran who operated a store on the northwest corner of State Routes 7 and 84. He died on his 93rd birthday in 1880, around the time this photograph of his son Harlay (seated in the center) and his family was taken. The home, built in 1823, is still is located on State Route 84 west of Bushnell corner. (Marianne Gullo.)

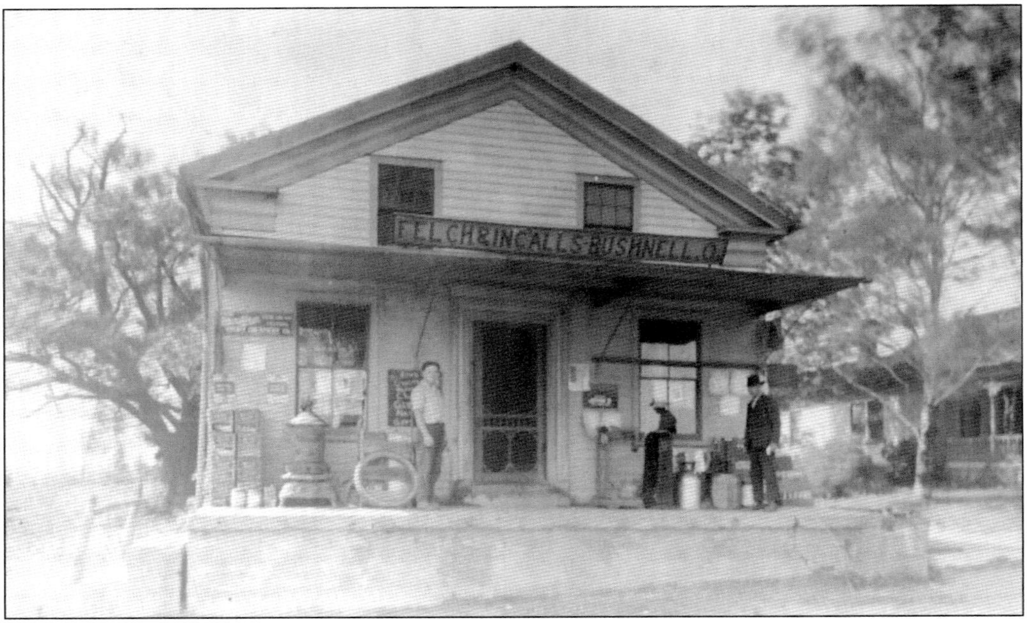

This store was operated by Sedgewick Bushnell until around 1845. It was later purchased by Irad P. Felch (on right) and Joseph R. Ingalls. Ingalls and Felch operated a post office, bank, and Ford dealership from the store. The store and a stagecoach inn on the southeast Bushnell corner were torn down to widen the intersection in the mid-1900s. M. M. Simmonds is to the left in this image. (Jennifer Krenisky.)

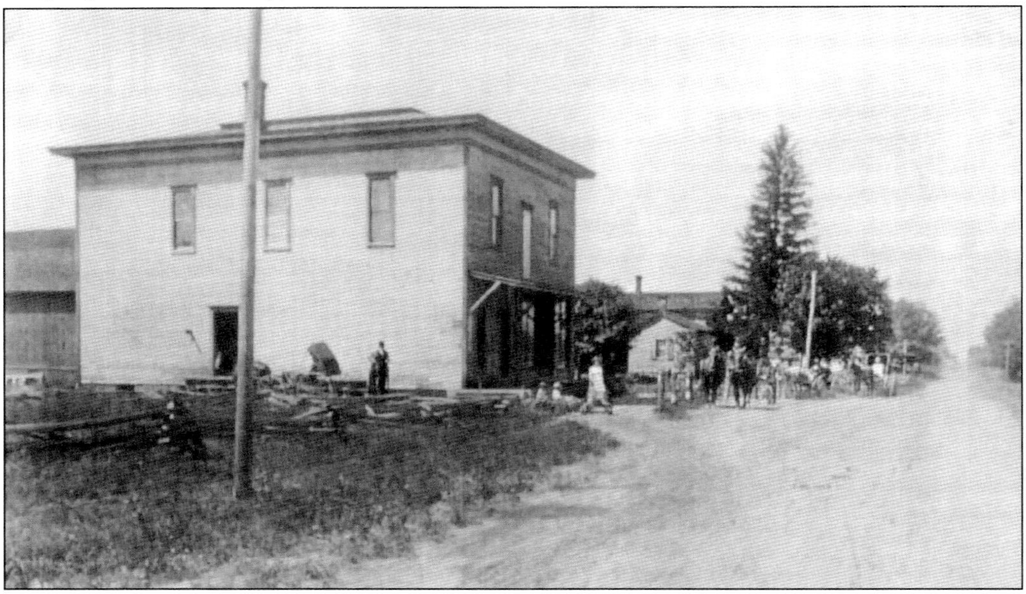

In 1868, George L. Felch and H. F. Hitchcock opened a store in this building south of Monroe Center Road on today's State Route 7. Hitchcock and Felch were purveyors of dry goods, notions, boots, shoes, groceries, and country produce. The store burned in 1915. In 1880, Felch married Kate Ingalls, who was a teacher. Kate's sister Laura married George's nephew, Irad P. Felch of Bushnell, in 1888. (Jennifer Krenisky.)

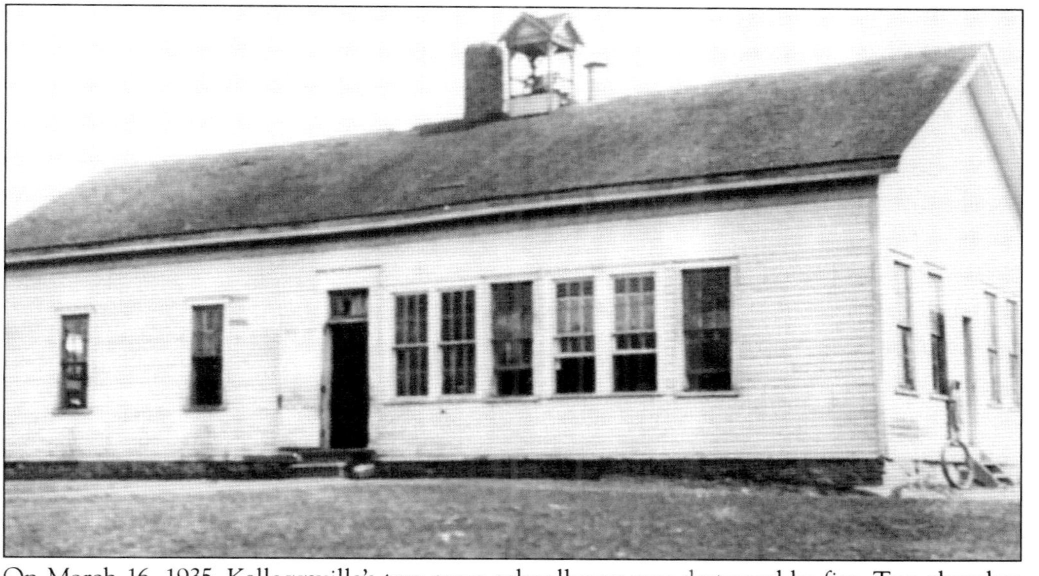

On March 16, 1935, Kelloggsville's two-room schoolhouse was destroyed by fire. Two churches provided classrooms for the remainder of the school year. In September 1936, the former Kelloggsville two-room schoolhouse students were transferred to the Kingsville School District. (Maurice Braunbeck family.)

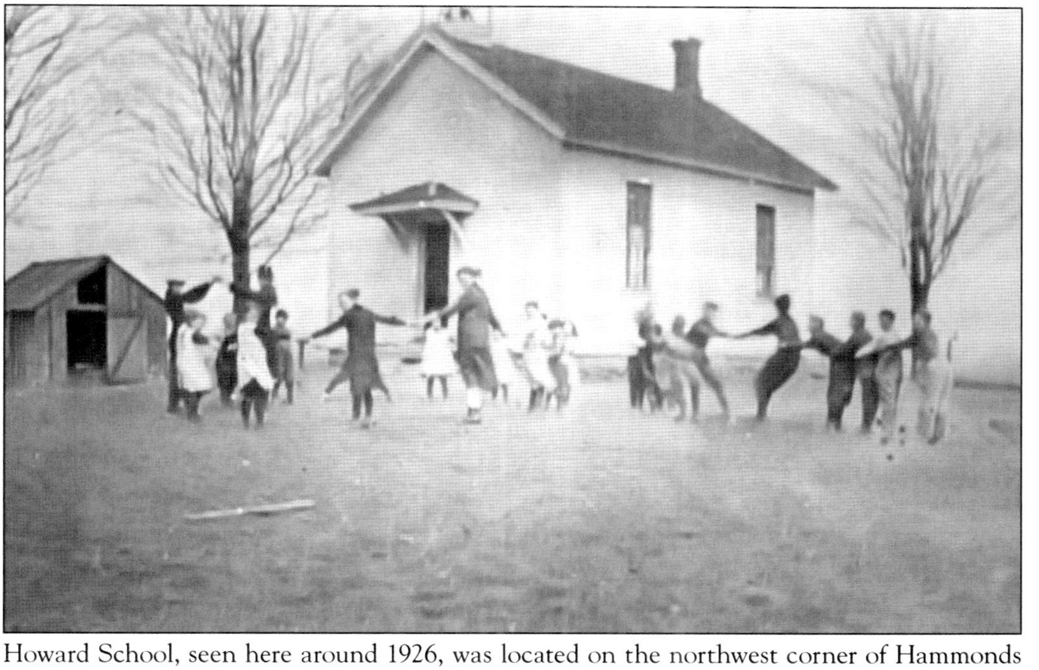

Howard School, seen here around 1926, was located on the northwest corner of Hammonds Corners and State Route 7. In 1937, this era ended when the Monroe Township School Board consolidated its six one-room schoolhouses into one building located on Monroe Center Road. This scene bears an eerie resemblance to Winslow Homer's *Crack the Whip* painting, which hangs in Youngtown's Ohio's Butler Institute of American Art. (Larry Carr.)

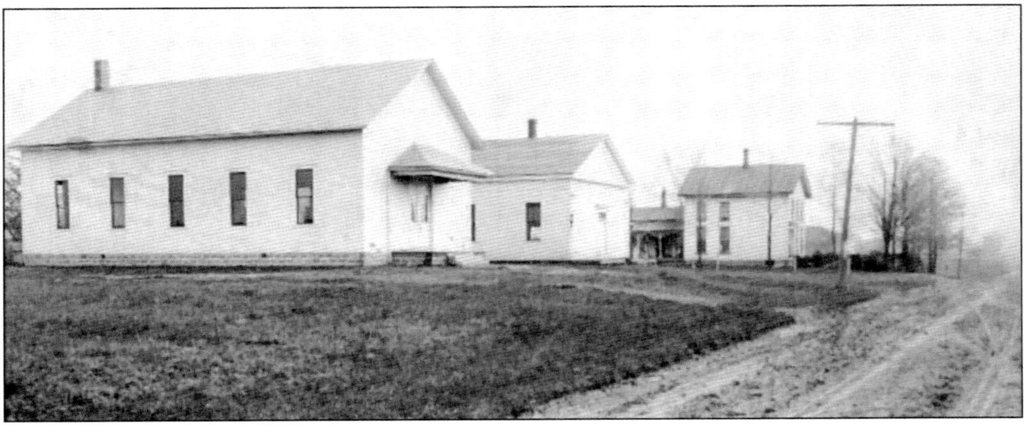

This 1911 photograph was taken looking southeast down an unpaved State Route 7 from the Monroe Center Road intersection. The first building on the left is the town hall. The second building is the Methodist church, which was built in 1852. A basement and bell would be added to the church over 100 years after it was built. (Jennifer Krenisky.)

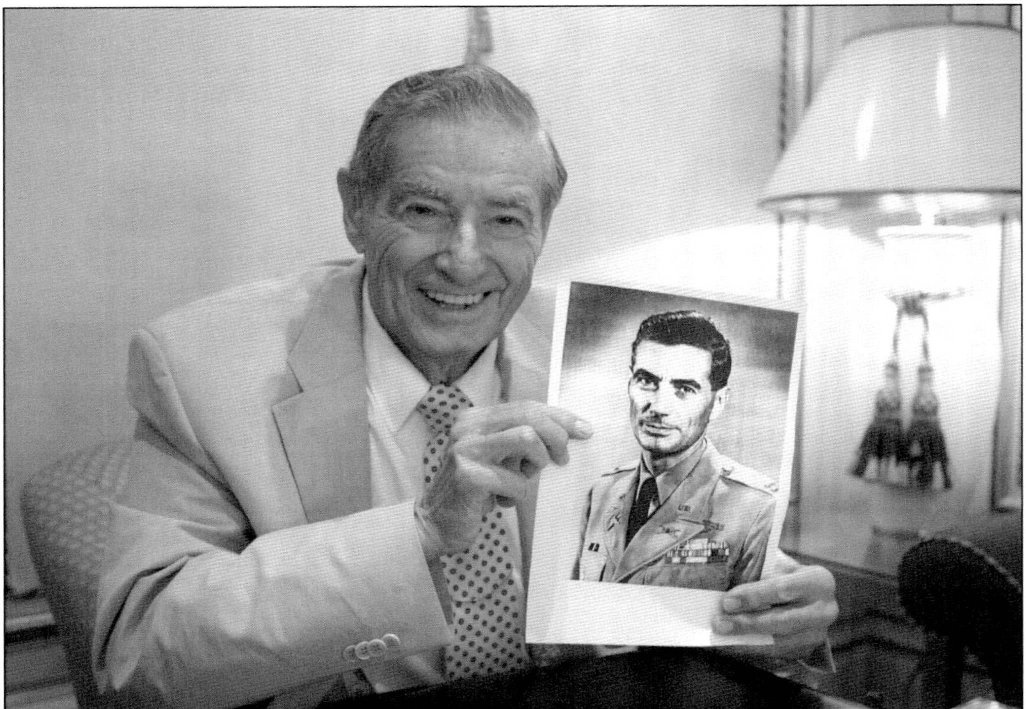

One of Monroe Township's favorite sons is Samuel W. Magill, who was born on Hildom Road in 1919. In September 1944, while Lieutenant Magill was in command of a U.S. Army reconnaissance patrol in central France, he successfully negotiated the surrender of Maj. Gen. Erich Elster and 20,000 troops. Thousands of American and German lives were saved without a single shot being fired. (Wolfgang Eilmes, Frankfurter Allgemeine Zeitung.)

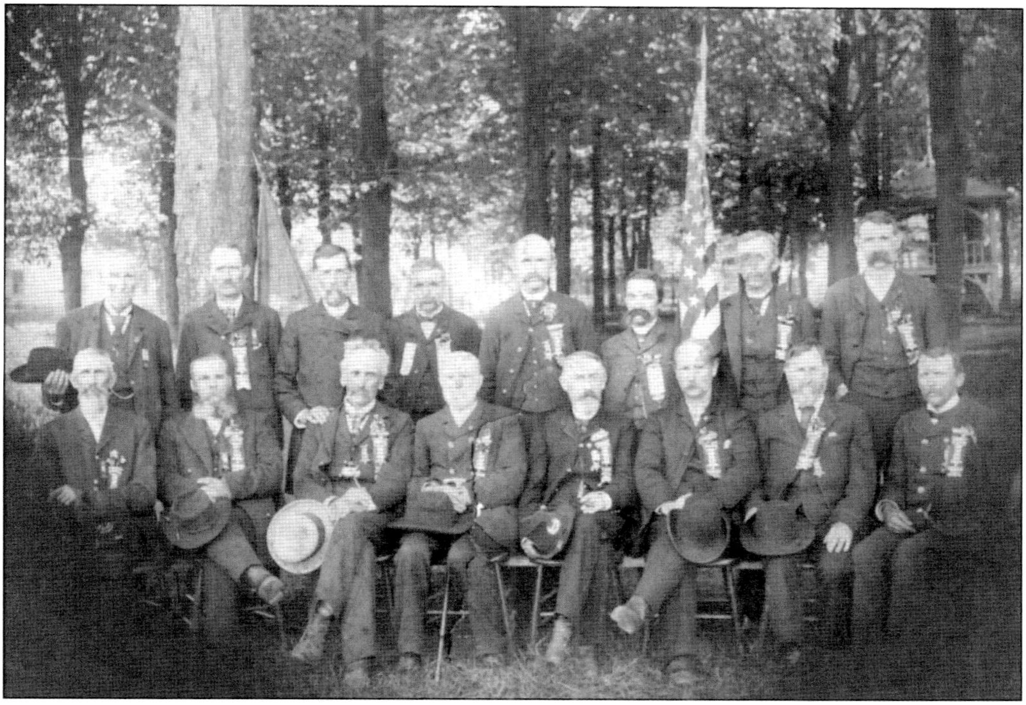

In this undated photograph of Grand Army of the Republic's Webster Post 8, one of Kingsville's leading citizens is seated fourth from the left. At age 44, Marshall William Wright volunteered to serve as quartermaster of the 105th Regiment Ohio Volunteer Infantry. After the war, he was elected Ashtabula County commissioner, served as Ashtabula County infirmary director, Kingsville Township trustee, Kingsville Academy trustee, and Kingsville postmaster. (Kingsville Public Library.)

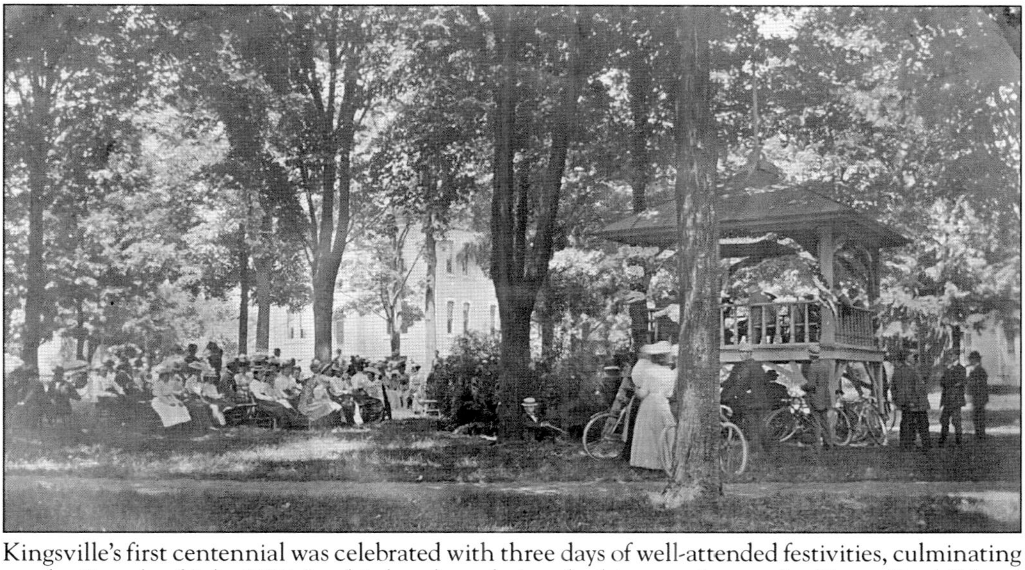

Kingsville's first centennial was celebrated with three days of well-attended festivities, culminating on the Fourth of July, 1905. In this bandstand view, looking northeast, the Centralized School on Academy Street can be seen beyond the audience. (Kingsville Public Library.)

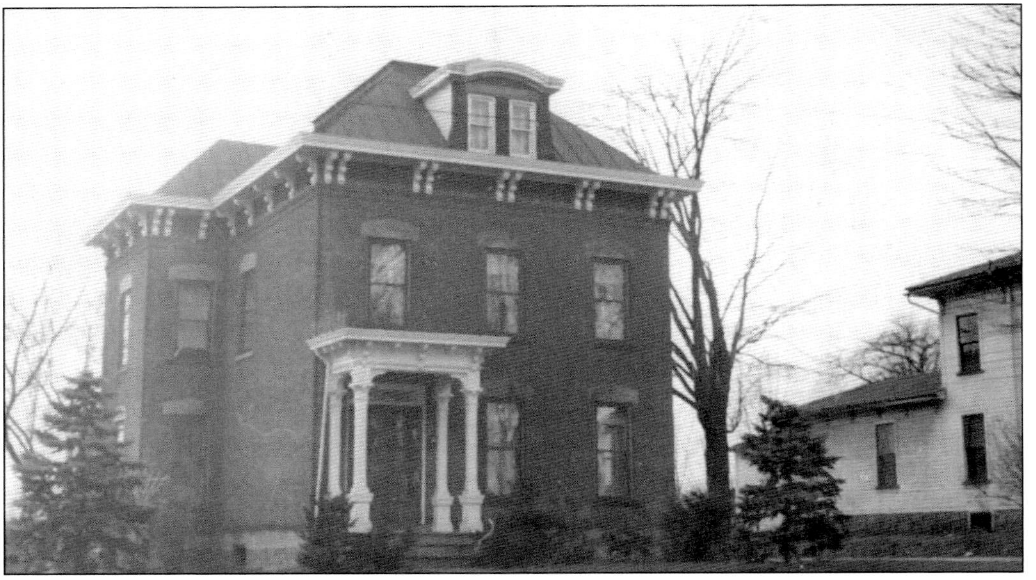

This home across from the park, on the west side of North Lake Street, is one of many Victorian-era homes that still grace the streets of Kingsville. It was built in 1882 by Edward F. Brown and his wife, Laura, who was the daughter of Jeremiah King. As a successful store owner, Brown and his wife were frequently able to spend winters in California. (Kingsville Public Library.)

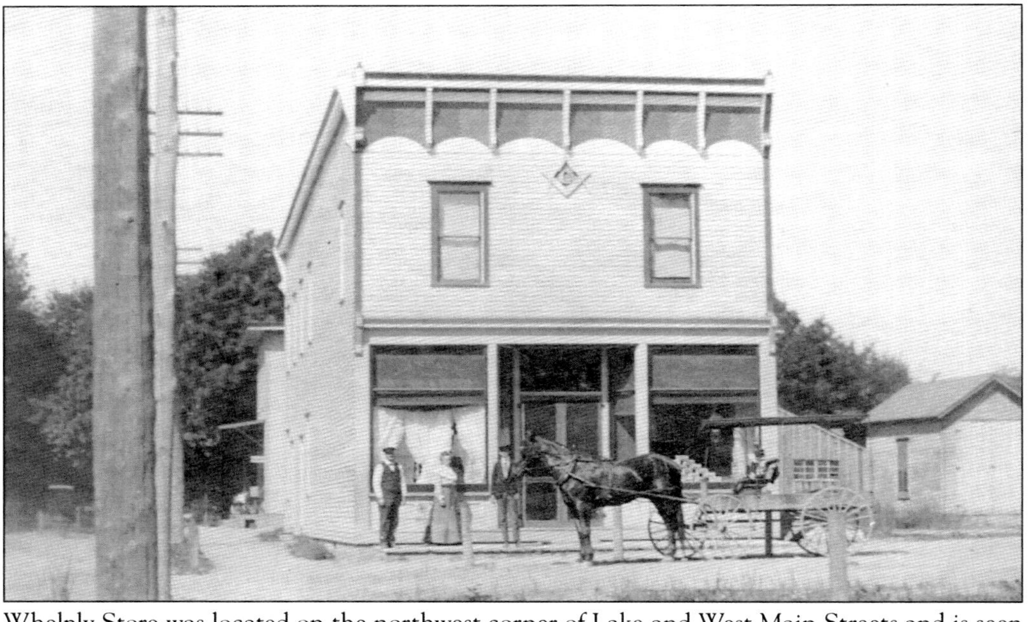

Whelply Store was located on the northwest corner of Lake and West Main Streets and is seen here around 1899. The building was owned by the Masonic Order. Jacob Kinnear's Hardware Store can be seen behind the Whelply Store with the Eastman barn to the right. (Kingsville Public Library.)

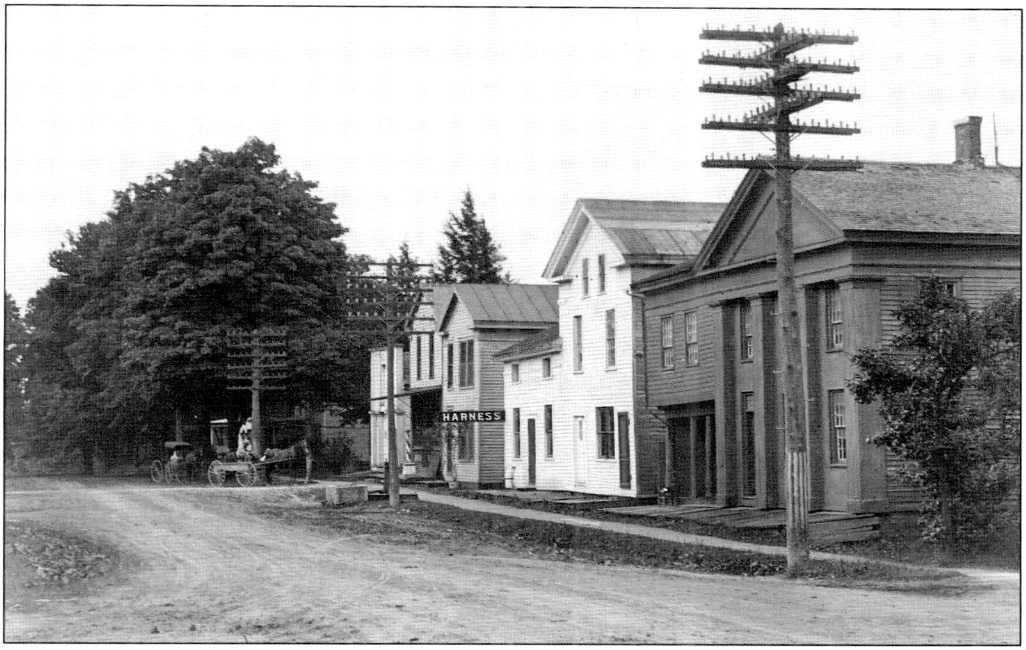

This *c.* 1900 view, looking east from Lake Street, is the south side of East Main Street. Buildings from right to left are the Rogers Hotel, Surline Hotel, Woodbury home and barbershop, Mullett Shop with apartments upstairs, the Grange Hall, and Gee Harness Shop. (Kingsville Public Library.)

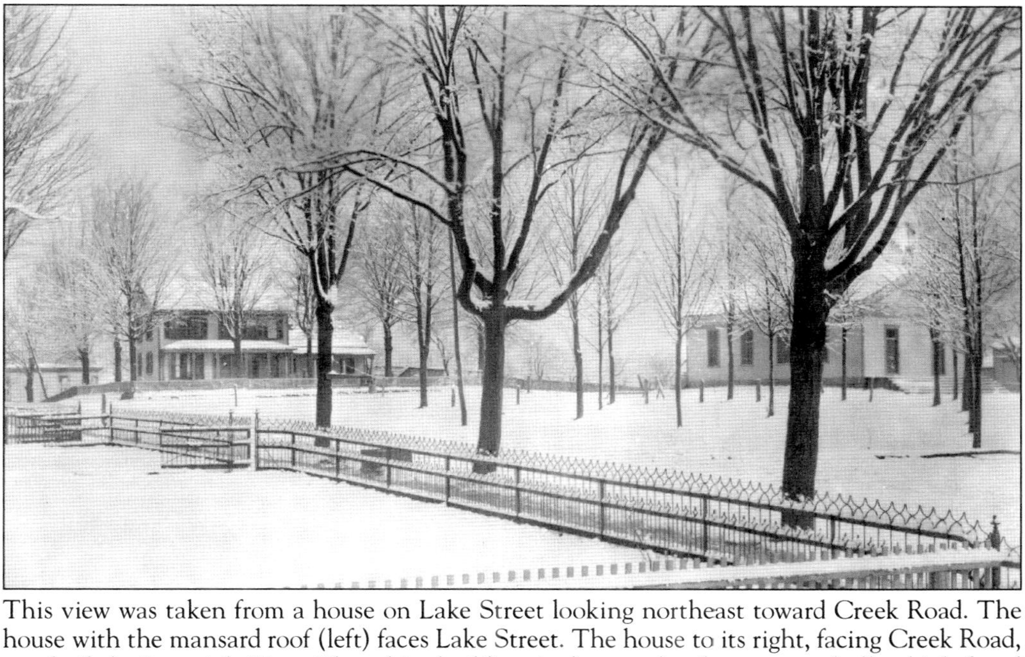

This view was taken from a house on Lake Street looking northeast toward Creek Road. The house with the mansard roof (left) faces Lake Street. The house to its right, facing Creek Road, was built by Jeremiah King. The white building is the wooden-frame town hall, which faced towards the north end of the park. (Kingsville Public Library.)

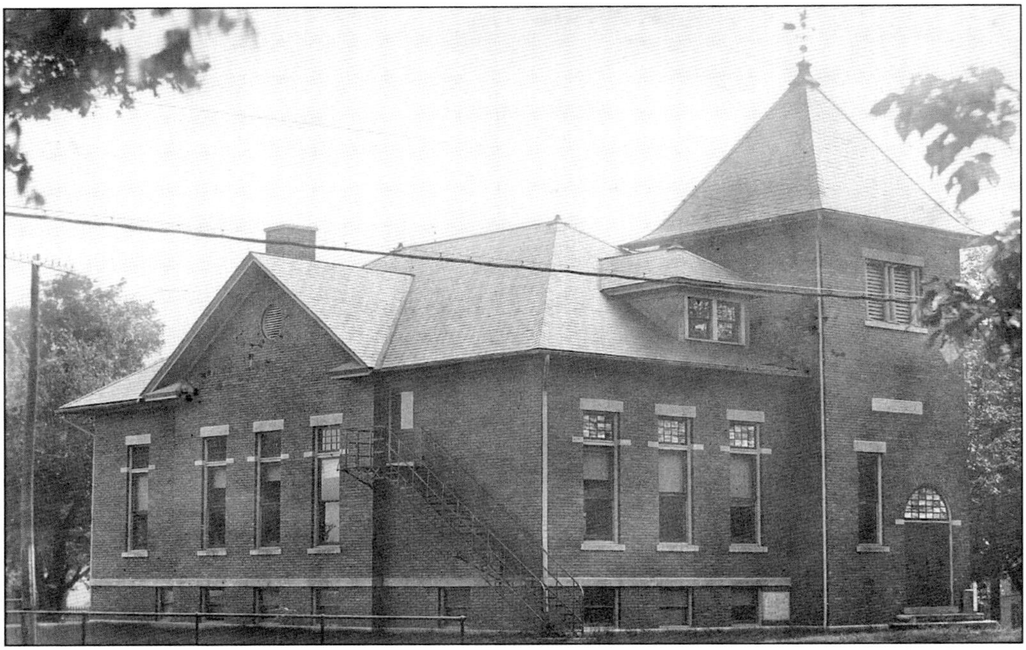

In 1911, the old wooden-frame town hall was moved north and west to make room for the new brick town hall built on the original location. This is a view of the Lake Street entrance from Creek Road. The building is now the Masonic Temple. (Kingsville Public Library.)

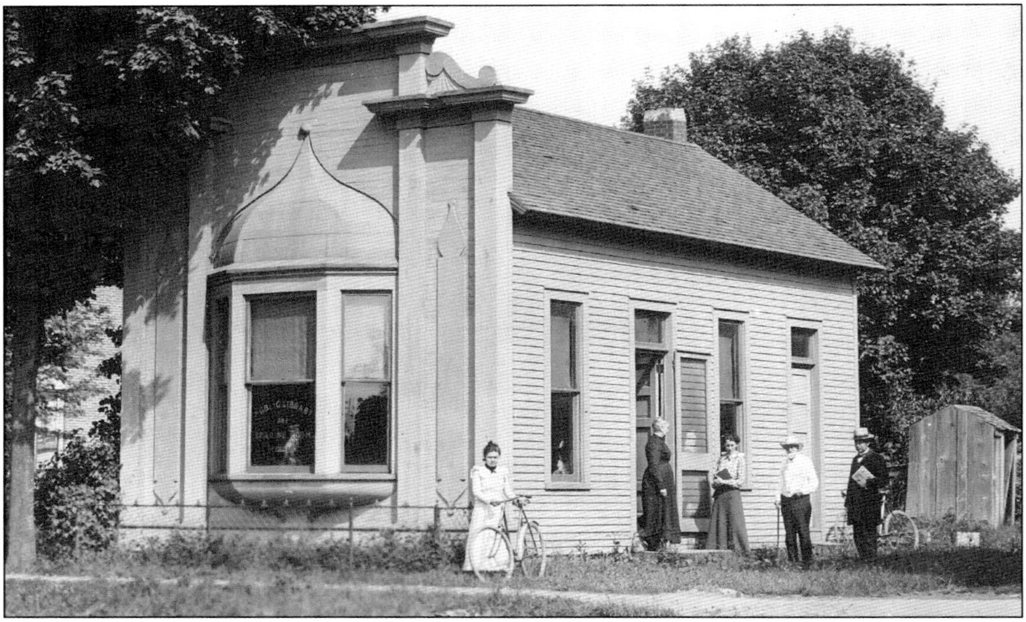

In 1885, the Kingsville Library Association was formed. In 1894, the Kingsville Public Library and Reading Room building was constructed at a cost of $609.61. In this 1905 photograph, the library is still in its original location on the north side of West Main Street. In 1918, the library building was moved to a site immediately south of the old Centralized School, where it sits today. (Kingsville Public Library.)

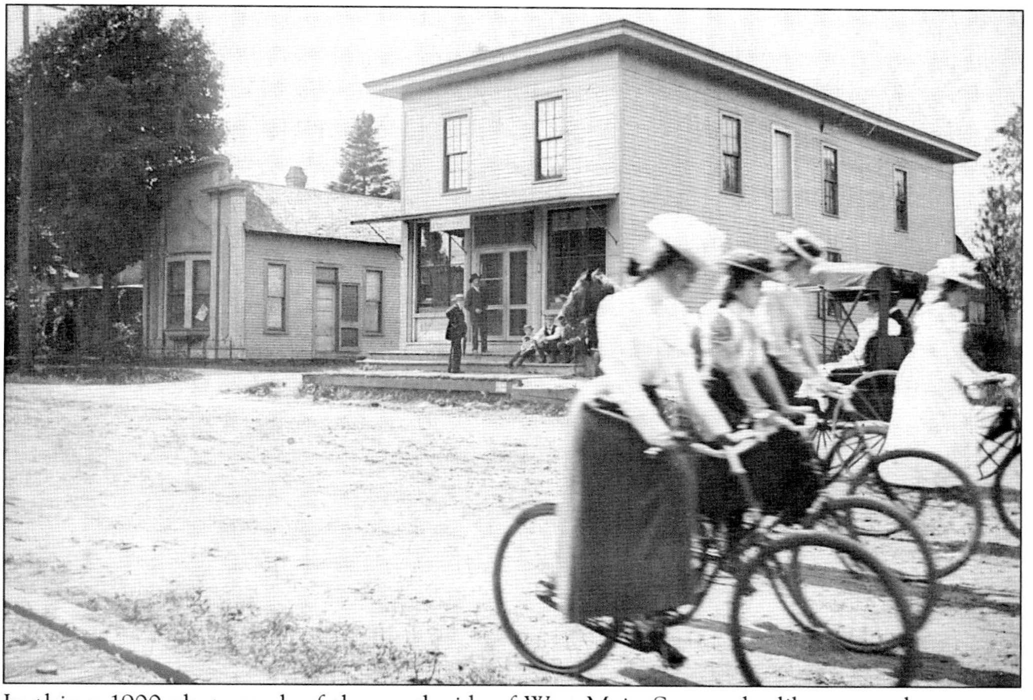

In this *c.* 1900 photograph of the north side of West Main Street, the library can be seen on its original site. The building in the center is Jacob Kinnear's Hardware, which opened around 1885. Thirty years before opening his store, Kinnear worked as a tinsmith with E. M. Curtis in a building connected with George Gillette's store. The building not visible to the right of Kinnear's Hardware is the Whelpley Store. (Kingsville Public Library.)

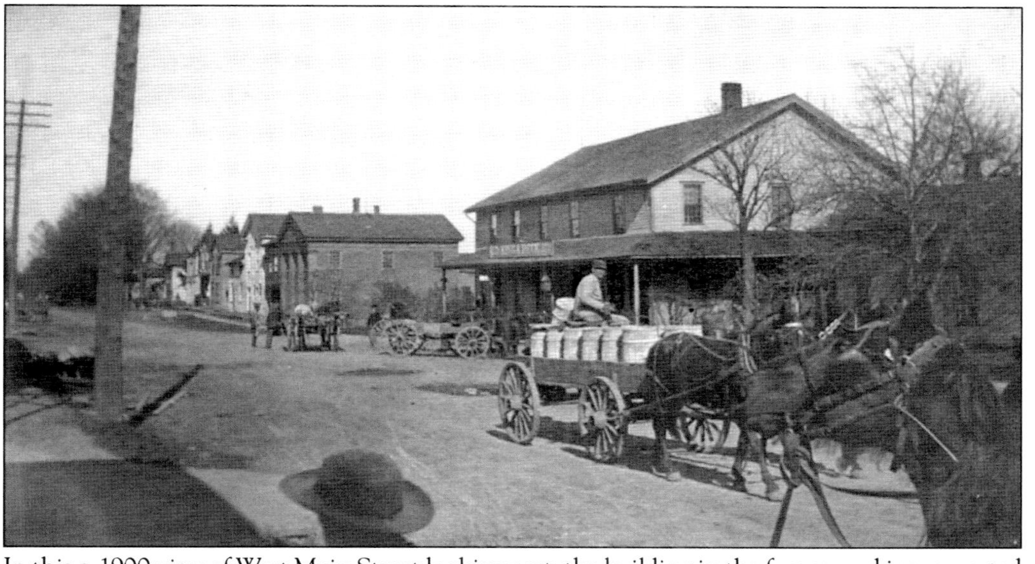

In this *c.* 1900 view of West Main Street looking east, the building in the foreground incorporated Kingsville's first frame building. George Gillette constructed the building in 1825 and added onto it as the business grew, including the post office in 1838. Gillette sold shares in the stock, and it was called the Farmers Store or Peoples Store. (Kingsville Public Library.)

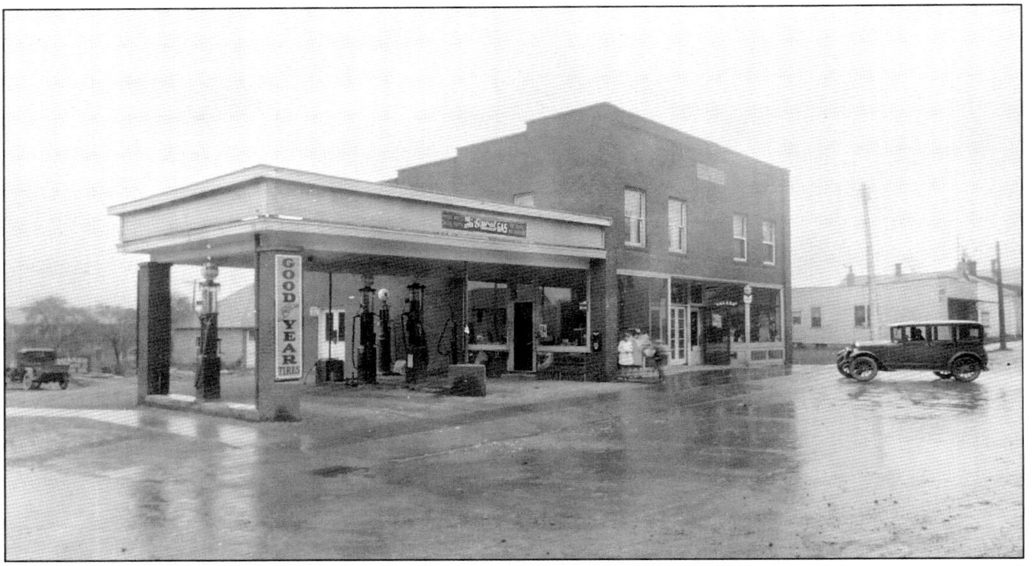

In this *c.* 1930 photograph, looking southwest from Lake Street, C. F. Tuttle's gas station and adjoining Tuttle Building can be seen on West Main Street. The Tuttle Building included stores on the ground floor and apartments on the upper floor. Today Kingsville's post office is housed on the first floor of the Tuttle Building, but the gas station is gone. (Kingsville Public Library.)

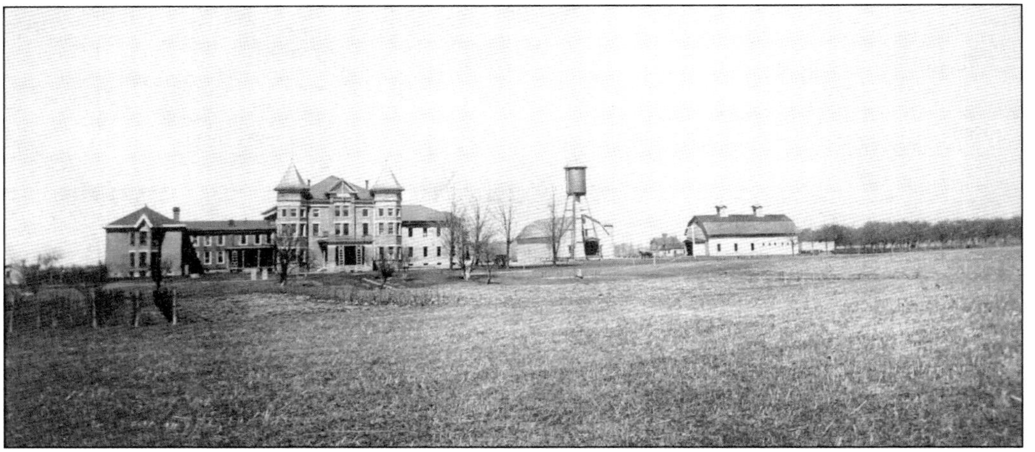

The Ashtabula County Infirmary was built in 1841 to house people who were unable to care for themselves. Census records indicate conditions of residents as old age, sickness, intemperance, blindness, deafness, dumbness, idiotic, insane, and simple among others. On February 2, 1858, a pipe-smoking patient set fire to her straw mattress, killing six residents. The surviving 60 residents were taken in by sympathetic Kingsville citizens until the infirmary was rebuilt. (Kingsville Public Library.)

William Whitney was born in 1842, becoming a resident of the county home and infirmary at the age of six and remaining there until he died at age 70. He acquired the nickname "Penny Billy" because of his curious habit of asking each visitor he encountered for a penny. The neckties he collected from people throughout the United States were made into a quilt by Carlotta Case. (Kingsville Public Library.)

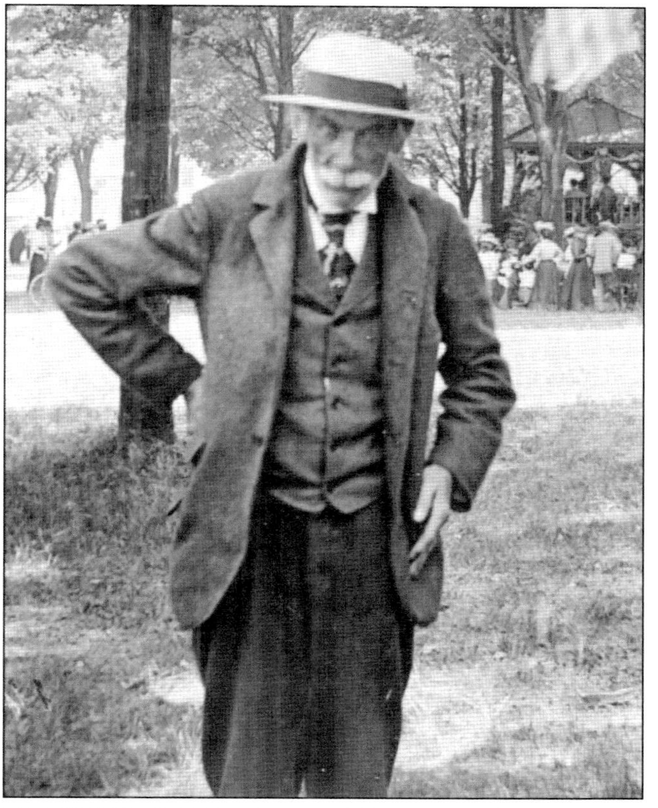

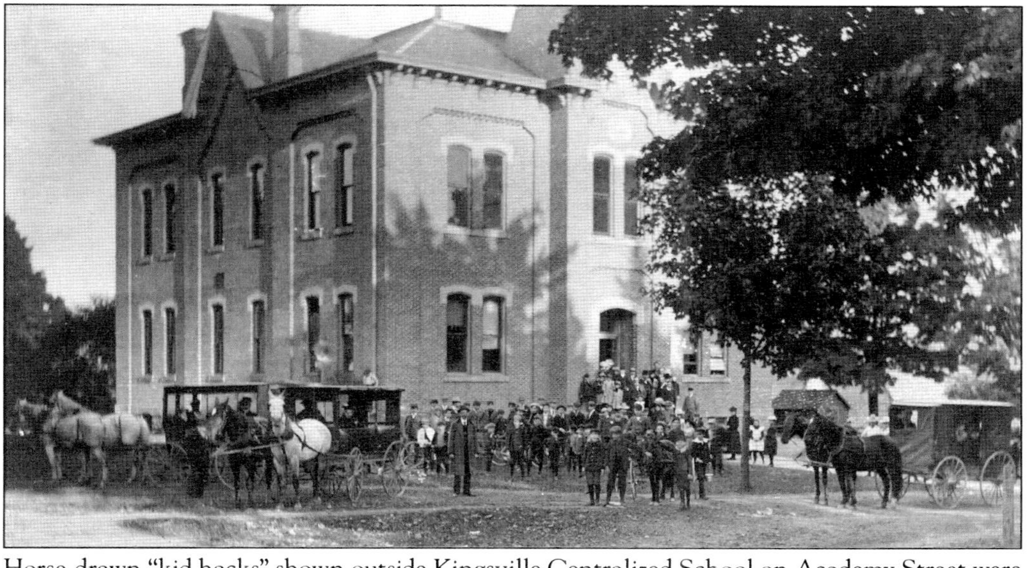

Horse-drawn "kid hacks" shown outside Kingsville Centralized School on Academy Street were Ohio's first busing program. The "Kingsville Plan," authorized by the Ohio State Legislature in 1892, reportedly saved Kingsville Township $1,000 the first year by consolidating sub-district schools. Additional benefits were improved attendance and the drivers doubling as mail carriers. (Kingsville Public Library.)

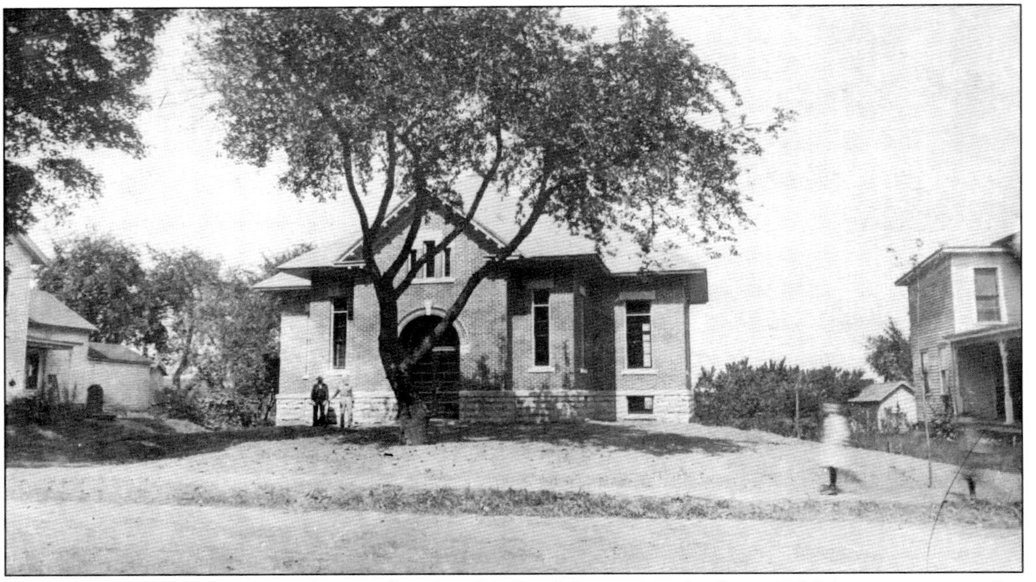

The Little Brick School on the north side of School Street was built in 1908 for primary grades to alleviate crowding in the Kingsville Centralized School. In 1963, the building was purchased by the Bugby Grange. The 3109 School Street home to the right was built in 1820. From 1915 to 1927, the home was used by the Ashtabula County Normal School as classrooms for teacher education. (Kingsville Public Library.)

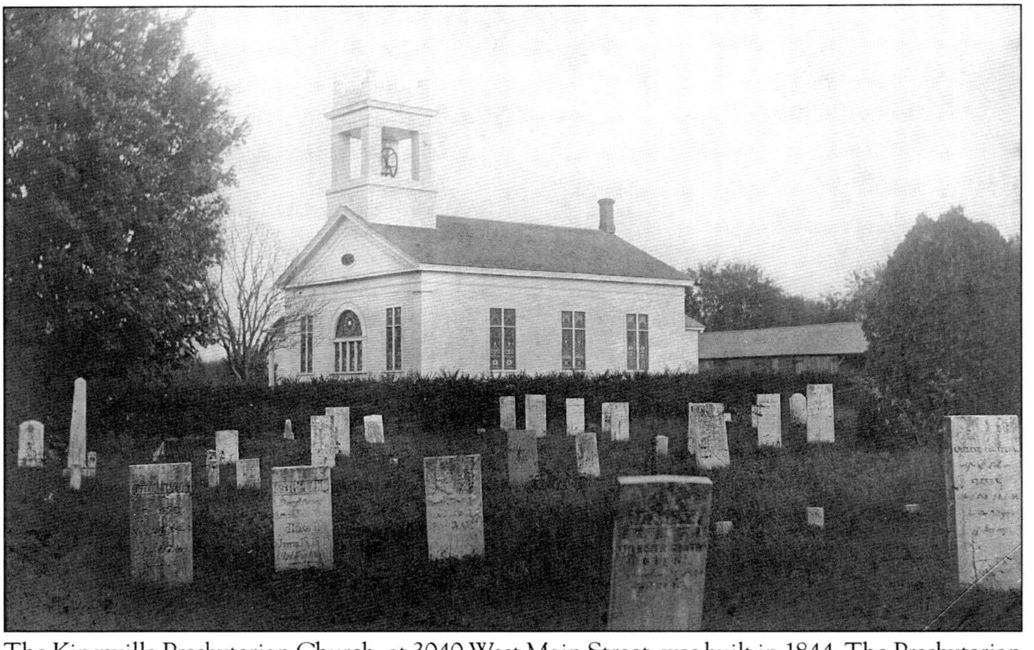

The Kingsville Presbyterian Church, at 3040 West Main Street, was built in 1844. The Presbyterian and Methodist churches merged in 1937 to form the Kingsville United Presbyterian Church. The old Methodist church building, on the opposite side of West Main Street, was moved in 1959 and now serves as the Presbyterian Fellowship Hall. The first burial in the Kingsville Cemetery (foreground) was in 1801. (Kingsville Public Library.)

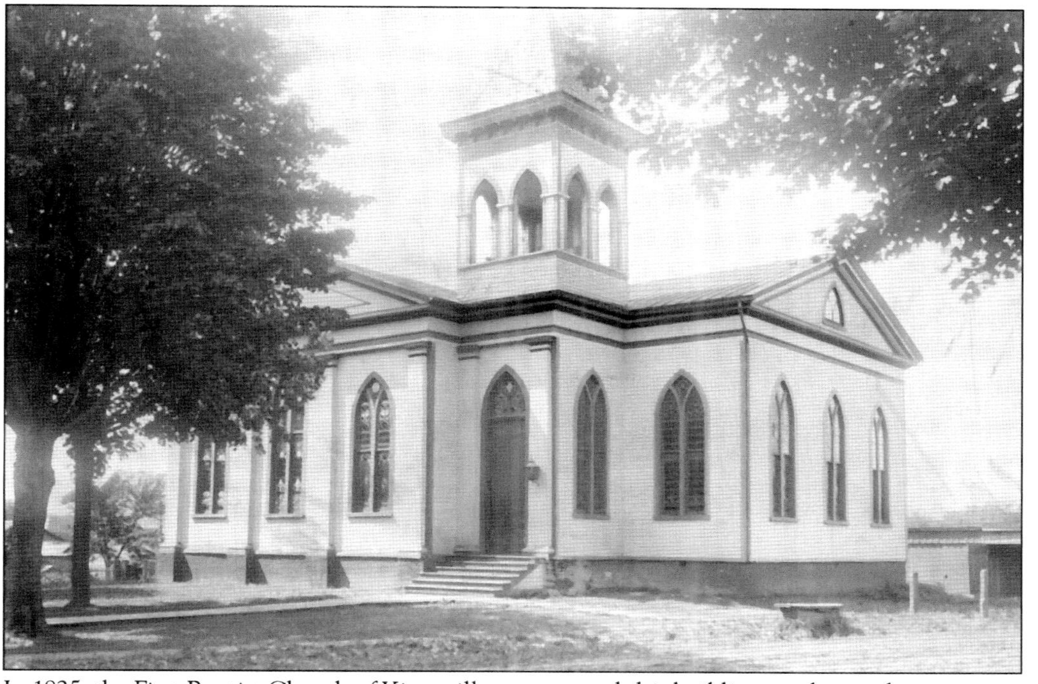

In 1825, the First Baptist Church of Kingsville constructed this building on the southwest corner of Lake and Priest Streets. Additions over the years included the chapel in 1894 and the bell in 1895. The church was destroyed by fire in 1909. Built on the same site, today's brick church was dedicated on February 7, 1911. (Kingsville Public Library.)

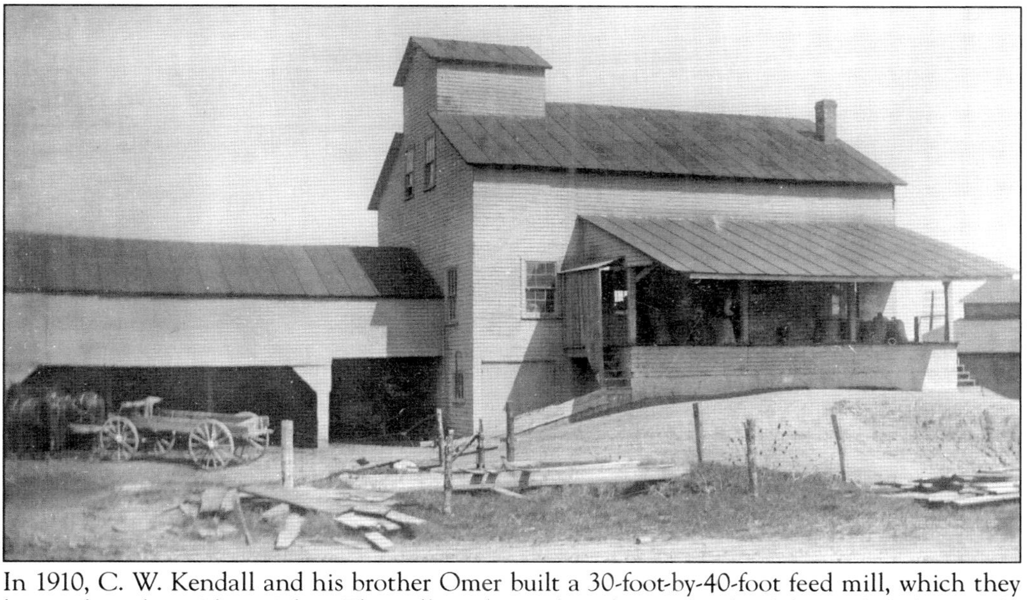

In 1910, C. W. Kendall and his brother Omer built a 30-foot-by-40-foot feed mill, which they later enlarged to 40 by 100 feet. The mill was located on the south side of the Nickel Plate tracks on the west side of Lake Street. Omer and his two children were killed when a train struck their automobile in 1923. In 1930, his brother sold the mill. (Kingsville Public Library.)

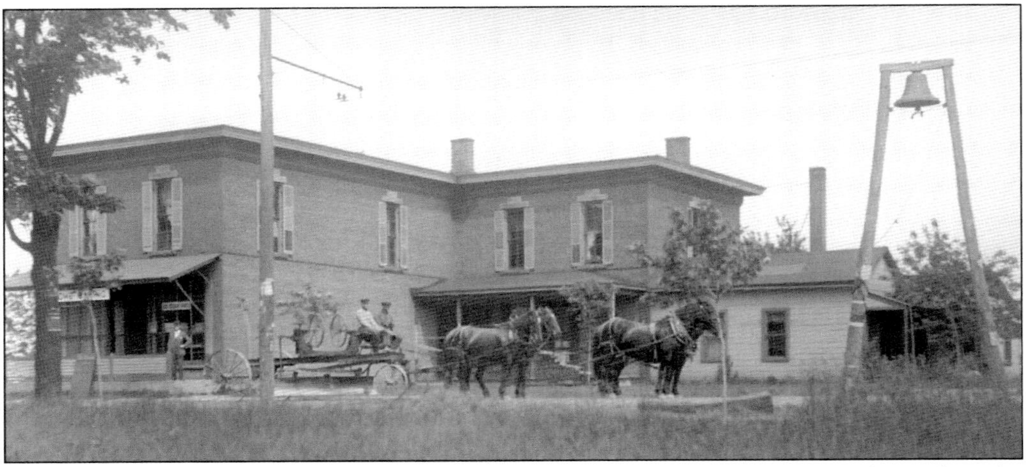

Lockwood Tavern featured a restaurant, rooms, and a mechanic's garage. It was located on the northeast corner of U.S. Route 20 and State Route 193 in North Kingsville. The Pennsylvania and Ohio Railway trolley would stop on the south side of Route 20 on its journey between Conneaut and Ashtabula. (Kingsville Public Library.)

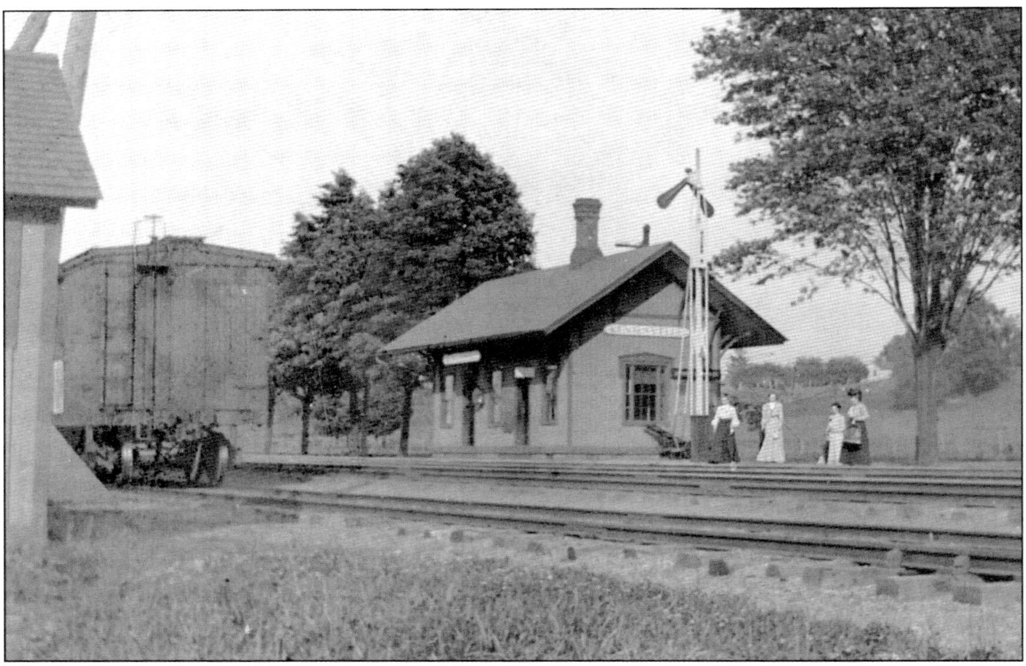

Kingsville Township's Lake Shore and Michigan Southern Railroad (later New York Central) passenger depot stood on the south side of the tracks on the east side of North Main Street. In this view, looking southeast, the approximate location of the house in the distance would be 945 East Center Street in North Kingsville. The building on the left is the freight house. (Kingsville Public Library.)

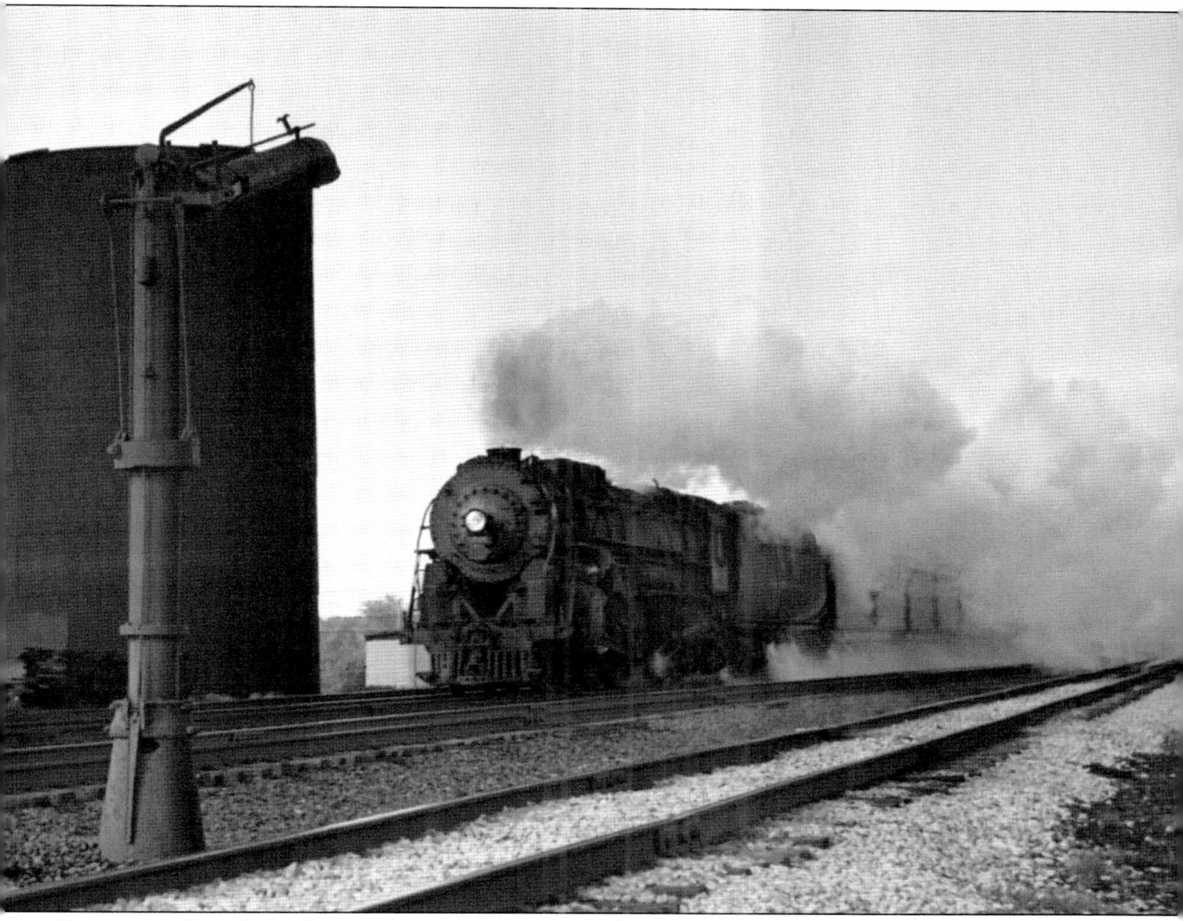

From this New York Central water tower in North Springfield, Pennsylvania, locomotives could spend several minutes ingesting water through the water crane in the foreground or run through the pans between the water crane and the tower. Specially fitted locomotives, traveling approximately 40 miles per hour, would lower a scoop into the water in the pans between the tracks, forcing thousands of gallons of water into their tenders in seconds.

Bibliography

Ashtabula County Genealogical Society. *Ashtabula County History; Then and Now; A History of the People of the County, By the People of the County*. Dallas, TX: Taylor Publishing Company, 1985.

Bawal, Raymond A. Jr. *Twilight of the Great Lakes Steamer, the Last Steam Powered Freighters on the Great Lakes*. Clinton Township, MI: Inland Expressions, 2009.

Biographical History of Northeastern Ohio: Embracing the Counties of Ashtabula, Geauga and Lake. Chicago: Lewis Publishing Company, 1893.

Erf, Edward. "The 'Kingsville Plan' of Education." *The Arena, Volume XXII, July to December 1899*: 61–68. New York: Alliance Publishing Company, 1899.

Harris, Chester. *Tiger at the Bar. The Life Story of Charles J. Margiotti*. Baumgratz Publishing, 1999.

Hilton, George W. *Great Lakes Car Ferries*. Montevallo Historical Press, Inc., 2003.

Kolehmainen, John I. *A History of the Finns in Ohio, Western Pennsylvania, and West Virginia*. Ohio Finnish-American Historical Society, 1977.

Smith, Joseph and Heman Conoman Smith, Francis Henry Edwards. *The History of the Reorganized Church of Jesus Christ of Latter Day Saints, Volume 4 (1873–1890)*. Independence, MO: Herald Publishing House, 1976.

Upton, Harriet Taylor. *The Western Reserve*. Lewis Publishing Company, 1910.

Watson and Dorman. *Directory of All Business and Professional Men of Ashtabula County Ohio*. Conneaut, OH: Rieg and Smith Publishing Company, 1895.

Welsh, Raymond. "Conneaut: The Story of Its Growth." *Conneaut News Herald*. July 2, 1946.

Williams, William W. *History of Ashtabula County, Ohio with Illustrations and Biographical Sketches of Its Pioneers and Most Prominent Men*. Philadelphia: Williams Brothers, 1878.

ABOUT THE ORGANIZATION

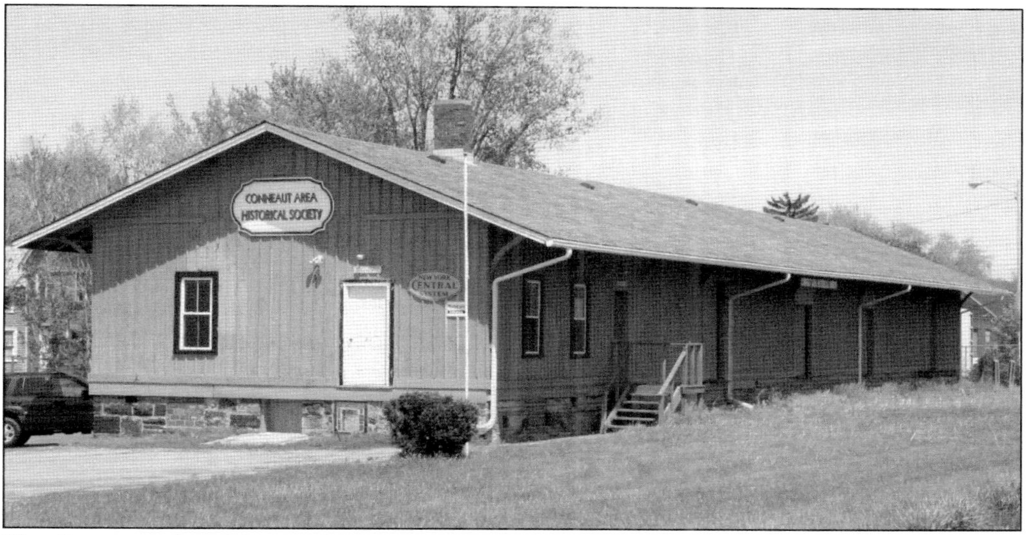

The Conneaut Area Historical Society was organized in November 1996 to collect and preserve artifacts and documentation of Conneaut area history. The society also strives to disseminate knowledge about the Conneaut area to others through its meetings, lectures, and museum exhibits.

The historical society's museum is located in the old New York Central Railroad freight depot next to the CSX Railway crossing on Mill Street. The 30-foot-by-140-foot building, home to the society's collection of artifacts and memorabilia related to Conneaut and neighboring communities, is open weekends from Memorial Day through Labor Day. The handicapped-accessible museum is home to a vast collection of photographs, artifacts, and memorabilia from Conneaut and surrounding communities.

Upon entering the west room entrance, one is greeted by many displays and a wall of photographs, which tell the story of local history. The middle room is devoted to research with volumes of information available for historical purposes. The east room of the building is devoted to larger displays. Many items are stored and rotated into the display room over the summer.

In 2006, the society acquired a famous architectural landmark, the David Cummins octagon house at 301 Liberty Street, and is in the process of restoring it to its once-elegant condition. The house was built in 1863 and is listed on the National Register of Historic Places. The historical society is in the process of furnishing and decorating the house true to the Civil War period.

The Conneaut Area Historical Society meets on the fourth Tuesday of each month at 7:00 p.m. at the Villa at the Lake at 48 Parrish Road in Conneaut. The society actively seeks historical documents, photographs, and artifacts to preserve Conneaut's heritage for the education and inspiration of future generations.